S0-AHD-516

Oxidative Stress, Exercise and Aging

DISCARDED

Oxidative Stress, Exercise and Aging

HELAINE M ALESSIO
ANN E HAGERMAN

Miami University, USA

EDITORS

Imperial College Press

Published by

Imperial College Press
57 Shelton Street
Covent Garden
London WC2H 9HE

Distributed by

World Scientific Publishing Co. Pte. Ltd.
5 Toh Tuck Link, Singapore 596224
USA office: 27 Warren Street, Suite 401-402, Hackensack, NJ 07601
UK office: 57 Shelton Street, Covent Garden, London WC2H 9HE

British Library Cataloguing-in-Publication Data
A catalogue record for this book is available from the British Library.

OXIDATIVE STRESS, EXERCISE AND AGING
Copyright © 2006 by Imperial College Press

All rights reserved. This book, or parts thereof, may not be reproduced in any form or by any means, electronic or mechanical, including photocopying, recording or any information storage and retrieval system now known or to be invented, without written permission from the Publisher.

For photocopying of material in this volume, please pay a copying fee through the Copyright Clearance Center, Inc., 222 Rosewood Drive, Danvers, MA 01923, USA. In this case permission to photocopy is not required from the publisher.

ISBN   1-86094-619-4

Printed in Singapore by World Scientific Printers (S) Pte Ltd

We dedicate this book to our families for their support and inspiration.

Helaine M. Alessio
Ann E. Hagerman

PREFACE

This book is designed for individuals interested in learning about exercise-induced oxidative stress and its effects as we age. It covers concepts that have been developed and debated since 1956 when Denham Harman first introduced the Free Radical Theory of Aging. Our understanding of radicals, antioxidants, and oxidative stress has improved due to advances in technology as we have changed from relying on by-products of radical-induced cell damage to capturing radical signals *in vivo* to measuring specific gene expressions known to be activated by radical-induced cell signals. Furthermore, our appreciation of the pleiotropic (radicals can be helpful and harmful) and hormetic (a low level of radicals benefits proper cell functioning) properties of reactive oxygen species (ROS) has been enhanced by exercise and aging research. Topics chosen for this book reflect a focus on exercise and its impact on oxidative stress, which is known to influence aging processes.

The first chapter explains reactive oxygen species (ROS) so that the reader can appreciate the biochemical details and understand the importance of redox cycling in biological systems. Endogenous defense against ROS and the actions of major antioxidants are also covered. Chapter 2 presents interesting, new, and provocative information about similar ROS pathways and consequences that plants and animals experience, including environment, temperature, nutrition, oxygen availability, and disease.

Chapter 3 addresses the exercise continuum with isometric contractions on one extreme and dynamic contractions on the other. Since most exercise movement fall someplace along the continuum, muscles are usually exposed to varying levels of metabolic and mechanical stress. Oxidative stress, in particular, can occur during muscle actions throughout the exercise continuum. The different oxidative stress mechanisms during isometric and dynamic muscle contractions ultimately yield ROS. The study of exercise, aging, and oxidative stress usually includes basic research that relies on animal models and applied research that uses human models. In fact, plants experience oxidative stress in similar ways as animals do. In Chapter 4,

ROS produced by skeletal muscle are explored from different perspectives: natural selection, animal size, animal species, muscle fiber type, and aging, Exercise can produce ROS across the exercise continuum and Chapter 5 explores mechanisms and specific types of muscle contractions, exercises, and sports in which ROS and antioxidant activities change. Over time, oxidative stress in skeletal muscle changes due to up and down regulation of antioxidants. Age-related changes in skeletal muscle antioxidant activities and their roles as cell signalers and in cell damage and repair are elucidated in Chapter 6. This is followed by Chapter 7, which focuses on age-related changes in ROS and its role in muscle dysfunction and sarcopenia.

Oxidative stress, exercise, and aging in cardiac muscle are described in Chapter 8. Heart and skeletal muscle share common oxidative stress mechanisms such as ischemia-reperfusion and calcium overload. Heart muscle has some unique protective courses of action, including heat shock proteins, for protection against ROS, and these are highlighted.

The final chapter explains how exercise and oxidative stress affect genetic expressions that regulate health and aging. The interaction between genes and environment is unfolding in greater detail with advances in microarray technology that allows for the simultaneous analyses of tens of thousands of genes at once.

H.M.A

CONTENTS

CHAPTER 1

CHEMISTRY OF REACTIVE OXYGEN SPECIES AND ANTIOXIDANTS

Dugald C.Close[1] and Ann E. Hagerman[2]
[1]University of Tasmania, Hobart, Australia;
[2]Miami University, Oxford, OH

1.1 What are Reactive Oxygen Species?

Reactive oxygen species (ROS) are compounds derived from molecular oxygen, O_2, by partial chemical reduction. ROS thus have "extra" electrons[a]. ROS include the familiar oxygen compound hydrogen peroxide, H_2O_2, produced when O_2 is reduced with two electrons, and reactive forms of oxygen including superoxide, O_2^{-}, and hydroxyl radical, $OH^{.}$. Complete reduction of O_2 by addition of four electrons yields 2 molecules of water, H_2O, a stable compound that is not an ROS (Simic, 1988).

When a species has an uneven number of electrons it is called a free radical. The unpaired electron which makes the radical unstable is indicated by the superscript ($^{.}$). Molecular oxygen can be reduced with one electron to yield superoxide (O_2^{-}), the first species produced in many biological oxidative cascades. Hydrogen peroxide can undergo homolytic cleavage to yield two hydroxyl radicals ($OH^{.}$). Free radicals are considered ROS even when they are not directly derived from oxygen. For example, oxidation of an unsaturated lipid by $OH^{.}$ yields the lipid radical and water [Eq. (1.1)].

---

[a] Reduction, the addition of electrons to one species, is always at the expense of oxidation, or removal of electrons, from another species.

$$\text{OH} \cdot \; + \; \text{R1} \diagdown \diagup \underset{\text{H}_2}{\text{C}} \diagdown \text{R2} \; \longrightarrow \; \text{H}_2\text{O} \; + \; \text{R1} \diagdown \diagup \underset{\text{H}}{\overset{\cdot}{\text{C}}} \diagdown \text{R2} \qquad (1.1)$$

Ground state O_2 is unusual in that it is a biradical, with two unpaired electrons. The relative stability of the ground state oxygen radical is because the unpaired electrons have identical spins (triplet state), making them unreactive with ordinary singlet state (electron pairs with opposite spin) compounds. Oxygen is reactive enough to be a useful substrate for the oxidative metabolism essential to life, but unreactive enough to comprise around 20% of the earth's atmosphere. Input of energy can reverse the spin of one of the unpaired electrons to produce an excited state of oxygen known as singlet oxygen ($^1O_2$). Singlet oxygen is more reactive than ground state oxygen, and is an important ROS in photosynthetic organisms since chlorophylls can facilitate photochemical excitation of oxygen to the singlet state.

The superoxide radical is only moderately reactive with most biological compounds, but in aqueous solution it rapidly reacts to form hydrogen peroxide [Eq. (1.2)]:

$$2\,O_2^- \; + \; 2\,H^+ \; \longrightarrow \; H_2O_2 \; + \; O_2 \qquad (1.2)$$

H_2O_2 is moderately reactive, has a relatively long half-life and can diffuse some distance from its site of production and across cell membranes (Vranová *et al.*, 2002).

Hydrogen peroxide reacts with reduced metal ions such as iron or copper to form the highly reactive hydroxyl radical (Fenton chemistry) [Eq. (1.3)].

$$H_2O_2 \; + \; Fe^{2+} \; \longrightarrow \; OH^{\cdot} \; + \; OH^- \; + \; Fe^{3+} \qquad (1.3)$$

In biological systems, the metal ion can be reduced by superoxide, ascorbic acid, or a variety of other reducing agents. The redox cycle involving repeated reduction of the metal ion, and continued production of hydroxyl radical, ensures that only catalytic amounts of metal are required to produce hydroxyl radical. Unlike superoxide and hydrogen peroxide, hydroxyl radical reacts rapidly with organic compounds.

All of the ROS (superoxide, hydrogen peroxide, and hydroxyl radical) are produced as a consequence of normal metabolism, and have roles as cell signaling molecules as well as in defense from invading micro-organisms. However, when any ROS is produced in uncontrolled amounts it can damage proteins, DNA, and lipids (Gutteridge and Halliwell, 1996). Reaction of hydroxyl radical with unsaturated lipids is the most familiar cascade of radical induced damage (Fig. 1.1). Reaction of radicals with proteins can lead to oxidation of reactive amino acid side chains, to protein crosslinking and denaturation, and to damage to other nearby proteins. Oxidation of DNA leads to strand breaks and release of oxidized bases, particularly 8-oxo-deoxy-guanidine.

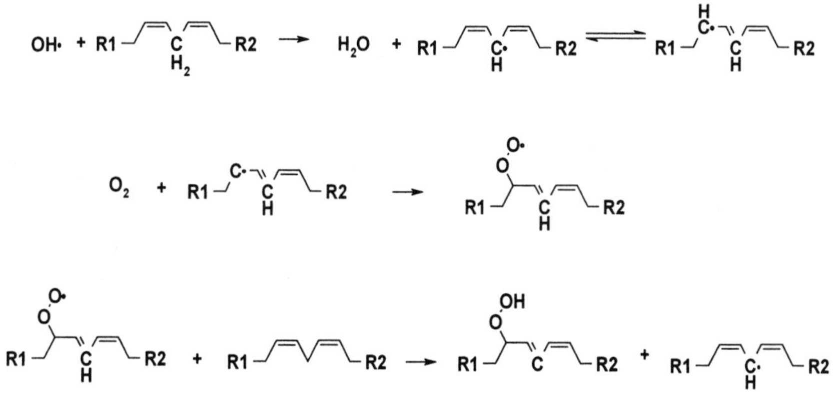

Fig. 1.1. Oxidation of unsaturated lipid to lipid hydroperoxide by OH·

Nitric oxide (NO·) is a biologically important radical produced by nitric oxide synthase (NOS) from arginine and oxygen. Nitric oxide is important as a cell signaling molecule, especially involved in vasoconstriction. As with other free radicals, uncontrolled production of nitric oxide can lead to oxidative damage.

1.2 What are Antioxidants?

Antioxidants minimize oxidative damage to biological systems either by preventing formation of ROS, or by quenching ROS before they can react with other biomolecules. Antioxidants can be either endogenous compounds, produced by the organism as part of its ROS defense, or can be exogenous compounds acquired from the diet. The endogenous system includes both enzymes and nonenzymatic antioxidants, and dietary antioxidants are small molecules.

All aerobic organisms contain the enzyme superoxide dismutase (SOD). Superoxide dismutase catalyzes the conversion of superoxide to hydrogen peroxide as shown above, providing about a 10,000-fold rate enhancement and ensuring that virtually no superoxide is found in the cell. In many tissues hydrogen peroxide is inactivated by catalase [Eq. (1.4)].

$$2\ H_2O_2 \xrightarrow{\text{catalase}} 2\ H_2O\ +\ O_2 \tag{1.4}$$

Glutathione peroxidase (GPX) provides an alternative route for destruction of hydrogen peroxide at the expense of the small molecule antioxidant glutathione (GSH). The oxidized glutathione (GSSG) is reduced via NADPH and glutathione reductase (GR). Analogous ascorbate peroxidases also play a role in hydrogen peroxide destruction [Eq.(1.5)].

$$2GSH\ +\ H_2O_2 \xrightarrow{\text{GPX}} GSSG\ +\ 2\,H_2O$$

$$GSSG\ +\ NADPH\ +\ H^+ \xrightarrow{\text{GR}} 2GSH\ +\ NADP^+ \tag{1.5}$$

Some antioxidants prevent ROS formation. For example, metal ion chelators such as transferrin and ceruloplasmin prevent metal ions from participating in Fenton chemistry, and thus minimize hydroxyl radical formation.

Small molecule antioxidants often act by scavenging, or quenching, free radicals. Glutathione and uric acid are important small molecule free radical quenchers found in plants and animals. In animals, the antioxidant vitamins (ascorbic acid, α-tocopherol and β-carotene) are radical scavengers. In plants ascorbic acid is endogenous, and is the most important small molecule antioxidant (Foyer, 1993). In the blood, non-specific radical quenching by albumins and other proteins contributes to total antioxidant capacity.

Glutathione is typical of several other thiol antioxidants including the amino acid cysteine and its derivatives, and thiol-containing proteins such as thioredoxins and peroxiredoxins. The thiol (sulfhydryl, R-SH) disulfide (R-S-S-R) pair provides a convenient redox couple used to destroy ROS; to regenerate other antioxidants; and to signal cellular redox status. Glutathione is a tripeptide (γ-Glu-Cys-Gly) with an unusual γ-glutamyl peptide bond, and a cysteine providing the thiol group (Fig. 1.2).

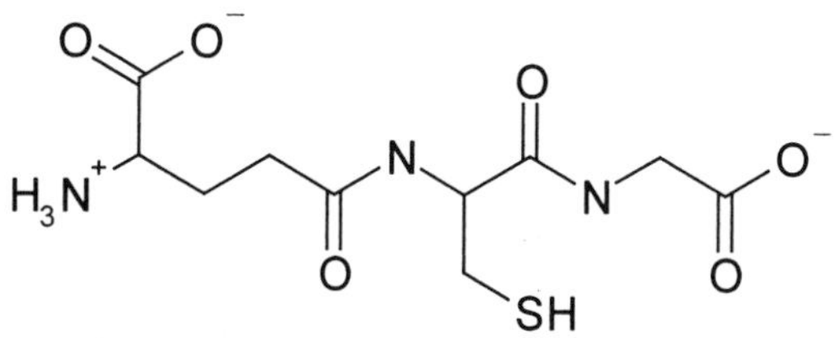

Fig. 1.2. Structure of glutathione.

Free radical scavenging is not always beneficial. For example, dietary polyphenols such as epigallocatechin gallate (EGCG), found in green tea react with radicals in a two step redox process. The EGCG is oxidized through the semiquinone radical to the quinone, while two radicals are reduced to nonradical forms. Both the EGCG semiquinone radical and the quinone are reactive species which can form covalent cross links to protein (Hagerman *et al.*, 2003). Although the original radical is destroyed by the "antioxidant" EGCG, oxidative damage may be promulgated by the altered polyphenol. For an antioxidant to be effective, it must not only quench radical species, but it must form relatively unreactive products that are not more damaging than the original radical.

Small molecule antioxidants often react in a network, involving multiple steps of oxidation/reduction to destroy the ROS (Blokhina *et al.*, 2002). One of the best characterized networks uses the fat-soluble vitamin α-tocopherol to protect membranes from damage, and ascorbic acid/GSH to regenerate the tocopherol (Fig. 1.3). Tocopherol is synthesized only in plants, and is found mainly in the chloroplast where it protects membranes from lipid peroxidation and scavenges singlet oxygen (Hess, 1993). Carotenoids, also found within chloroplasts, also protect plants from singlet oxygen. The xanthophyll cycle is a redox cycle using specialized carotenoids to compensate for photo-oxidative stress encountered when plants are exposed to low temperatures and bright lights.

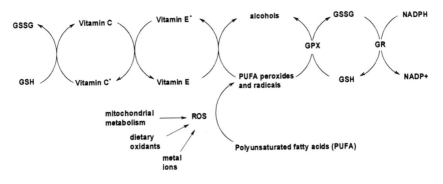

Fig. 1.3. Reduction of polyunsaturated fatty acid hydroperoxides (PUFA peroxides) and radicals by vitamins E, C, and GSH.

1.3 Oxidized Biomarkers

Biological molecules that are susceptible to oxidation damage include protein, lipids and nucleic acids. Selecting appropriate biomarkers to quantitate damage is challenging, because the amounts of oxidized products may be small; oxidized products are intrinsically unstable; and biologically significant targets may not have been identified for a given tissue. Common methods for measuring damage to protein include determination of protein carbonyls using direct chemistry or immunoassay. This is a nonselective endpoint, and better results may be obtained by determining specific modified amino acids such as nitrotyrosine. Oxidized lipids are indirectly determined as aldehydes such as malonyldialdehyde (MDA) using either simple colorimetric methods including thiobarbituric acid reactivity (TBARS assay) or more selective HPLC methods. MDA and related methods are indirect as the aldehyde is a degradation product of the oxidized lipid. More direct methods include direct spectrophotometric measurement of lipid hydroperoides, or determination of specific oxidized lipids such as isoprostanes. Oxidized nucleic acids are determined as metabolites such as hydroxypurines or pyrimidines, or as xanthine. Radical species themselves are very difficult to detect in biological samples because of their short lifetimes, but radicals can be spin trapped by reagents such as PBN for direct determination by electron paramagnetic resonance (EPR) or can be determined less directly by reaction with a fluorescent dye.

In addition to determining oxidized biomarkers, antioxidants are often estimated in biological samples. Total antioxidant activity can be measured by a wide variety of approaches ranging from the ORAC (oxygen radical absorbant capacity) to TEAC (trolox equivalent antioxidant capacity) methods. The antioxidant status of a tissue can be estimated by determining the ratio of oxidized to reduced GSH (GSSG: GSH). Individual antioxidants such as the vitamins can be directly determined by HPLC.

Before any of these measurements are attempted, investigators should explore the extensive literature available on methods for measuring oxidized biomarkers so that the most appropriate methods can be used for a given study (Armstrong, 2002).

1.4 Summary

The study of reactive oxygen species encompasses a wide range of compounds ranging from very reactive free radicals to the damaged biomolecules produced by uncontrolled oxidative stress. Understanding the chemistry of these species forms a basis for understanding the role of oxidative damage in disease, aging and exercise, as discussed in the following chapters of this book.

References

1. Armstrong, D. *Oxidative stress biomarkers and antioxidant protocols.* (Totowa, NJ Humana Press, 2002).
2. Blokhina, O.B. *et al., Ann. Botany* **91** (2002), 179-194.
3. Foyer, C.H. *Antioxidants in higher plants.* ed. Alscher, R.G. and Hess, J.L (Boca Raton: CRC Press, 1993), 31-58.
4. Gutteridge, J.M.C. and Halliwell, B. *Antioxidants and nutrition, health and disease.* (Oxford, Oxford University Press, 1996).
5. Hagerman, A.E. *et al., Arch. Biochem. Biophys.* **414** (2003), 115-120.
6. Hess, J.L. *Antioxidants in higher plants.* ed. Alscher, R.G. and Hess, J.L. (Boca Raton: CRC Press, 1993), 112-133.
7. Simic, M.G. *Oxygen radicals in biology and medicine.* (New York, Plenum Press, 1988).
8. Vranová, E. *et al., J. Exper. Bot.* **53** (2002), 1227-1236.

CHAPTER 2

OXIDATIVE STRESS IN PLANTS AND ANIMALS

Dugald C.Close[1] and Ann E. Hagerman[2]
[1]University of Tasmania, Hobart, Australia; [2]Miami University, Oxford, OH

2.1 Introduction

Nearly 50 years ago Denham Harman (1958) postulated that aging is caused by cell damage from rogue molecules that he referred to as free radicals. Harman recognized the reactive nature of free radicals and related forms of activated oxygen that we now call Reactive Oxygen Species (ROS). He proposed that ROS have played an important role in evolution of life on earth (Harman, 1981). Although early life forms developed in the absence of an oxygen atmosphere, the ability to deal with ROS had to develop in parallel with oxygen-dependent photosynthesis. As today's oxygen-containing atmosphere developed, further abilities to deal with damage due to ROS must have been introduced.

The time line (Fig. 2.1) highlights some major events over the past 3.5 billion years that include the change from radiation to oxygen-centered radical reactions in the atmosphere, the movement of living organisms from water to land, and the development of protective systems in aerobic organisms to exercise and diet supplementation to enhance endogenous antioxidant defense systems. The major biochemical pathways leading to production of ROS in animals and plants are outlined in this chapter. Conditions that may alleviate or enhance oxidative stress are described, and some of the antioxidant systems employed by animals and plants are listed.

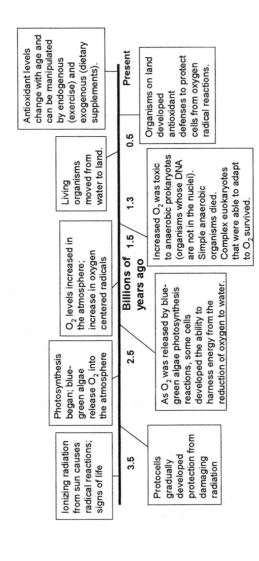

Fig.2.1. Time line for major events related to oxidative stress over 3.5 billion years.

2.2 Sources of Oxidative Stress in Plants

In plants, the highly energetic electron transport reactions of photosynthesis in chloroplasts and of respiration in mitochondria make these organelles major contributors of ROS. Excited chlorophyll can yield singlet oxygen species when there are inadequate electron acceptors available. Photosynthetic (Asada, 1999), mitochondrial (Purvis, 1997), peroxisomal and glyoxisomal (Tolbert, 1981) electron transport chains can become over-reduced when the enzymatic stages of photosynthesis lag behind the electron transport processes, ultimately generating ROS. NADPH oxidase in the plasma membrane, glycolate and xanthine oxidases, fatty acid β-oxidation in the peroxisome, and oxalate and amine oxidase in the apoplast (reviewed in Mittler, 2002) are other metabolic sources of oxidizing species in plant tissues.

2.3 Sources of Oxidative Stress in Animals

In animals, "leaky" mitochondrial electron transport is the major source of oxidative stress, with perhaps 2 to 5% of the electron flow generating superoxide and ultimately hydrogen peroxide (Grisham, 1992). Similar to plants, NADPH oxidase, amine oxidases and xanthine oxidase produce ROS. In addition, the microsomal mixed function oxidases may produce ROS as by-products of xenobiotic detoxification. Myoglobin and other heme-proteins have been implicated in radical production. For example, neutrophils and monocytes produce hypochlorous acid, a potent oxidizing agent, via a heme-containing myeloperoxidase.

2.4 Similar Consequences of Oxidative Stress for Plants and Animals

Under favorable conditions for growth, plants and animals strictly regulate the formation of ROS through a coordination of enzymatic and non-enzymatic antioxidants, described in the previous chapter. Under conditions of stress the production of ROS can increase dramatically (Polle, 2001). The relative severity of stress depends upon a myriad of factors including species and individual differences, but in any case severe stress can generate sufficient ROS to exceed the protective capacity of the antioxidant systems. In plants, damage to the membrane structure of organelles, to enzyme function and carotenoid pigments can cause death in extreme cases (Wise, 1995).

In both plants and animals, production of ROS is enhanced by biotic or abiotic stresses as described below. Oxidative damage can also be a consequence of loss of control over the normal inflammatory response, leading to auto-inflammatory diseases such as arthritis and inflammatory bowel disease. Finally, products of oxidative damage accumulate with age and may contribute to the characteristic degeneration of many animal physiological systems with age. Aging is not as pronounced in plants as in animals, and some higher plants live thousands of years. In plant and animal cells oxidases are tightly regulated to produce ROS during programmed cell death since even in long-lived species individual cells have limited life spans and are replaced by new cells.

2.5 Environmental Factors Inducing Oxidative Stress

Environmental stresses can cause metabolic changes in plants or animals that either increase production of ROS, or decrease production of antioxidants. Animals can sometimes avoid these stresses by choosing a new environment, but plants are constantly exposed to uncontrolled environmental stresses because they are immobile. Stresses that contribute to oxidative damage in plants or animals include excess radiant energy, temperature stress, pollutants or toxic chemicals, nutrient stress, competing organisms including pathogens and parasites, and anoxia.

2.6 Irradiation

UV light provides relatively high energy inputs to unscreened organisms on earth. In general, UV light is divided into UV_A (320–400 nm), UV_B (280–320 nm) and UV_C (180–280nm). Of these UV_B is most relevant, as UV_C does not reach the earth's surface and UV_A is far less energetic. UV_B radiation can cause damage to DNA, proteins and membranes

(Jordan, 1996), and contributes to skin cancers and cataracts in humans. Many organisms are well-protected from UV$_B$ radiation, via dermal modifications such as pigmentation and hair in animals. Similarly, the flavonoids synthesised in the epidermis of higher plants effectively shield plants and prevent damage from UV$_B$ radiation (Landry *et al.*, 1995). It is not clear whether significant increased damage will accompany the increased levels of UV$_B$ radiation expected as a consequence of depletion of the stratospheric ozone layer.

High energy irradiation such as gamma irradiation and X-rays do cause oxidative damage, but are very minor contributors to overall oxidative stress except during certain medical treatments, or in environments with high natural levels of radioactivity.

The fundamental photosynthetic process in plants is designed to absorb visible light and use it to excite electrons through an electron transport chain (Fig. 2.2). Oxidative stress in plants under conditions of extreme irradiation is common, although the intensity of stress is species and condition dependent. Typical conditions promoting light-induced oxidative stress are bright light and low temperatures, when the ability of the photosynthetic apparatus to excite electrons exceeds the capacity of the protein-driven electron transport chain and dark reactions to use the excited electrons. Garcia-Plazaola *et al.* (2004) reported that ascorbate, glutathione and α-tocopherol pools increased with increasing irradiance through a natural forest canopy, but that only α-tocopherol responded to extra irradiance. Consistent with a general response of plants of higher antioxidant content with higher incident visible radiation, Ma and Cheng (2003) showed that the peel of sun-exposed sides of apples had higher activities of ascorbate peroxidase and glutathione reductase and higher size and higher radiation state of the ascorbate pool and the glutathione pool. Ma and Cheng (2004) subsequently showed that exposure of the shaded side of apples to full sun led to acclimation of the antioxidant component to be similar to the sun-exposed controls by 10 days. Visible light does not cause severe damage to animals, although frequent exposure to very bright visible light can affect the ability of the eye to dark adapt.

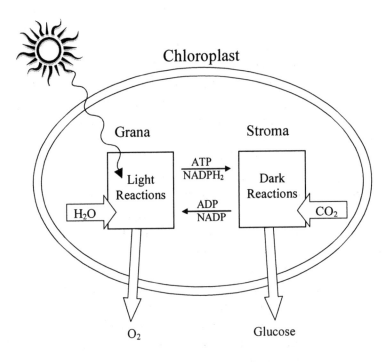

Fig. 2.2. Schematic of light and dark reactions of photosynthesis.

2.7 Temperature Extremes

Heat shock detrimentally affects photosynthetic and oxidative electron transport and alters the integration of enzymatic and antioxidant molecules (Paolucci *et al.*, 1997) and thus leads to the formation of ROS (Ando *et al.*, 1997; Gong *et al.*, 2001). Accumulation of lipid hydroperoxides in the liver is a consequence of heat stress in animals, and organelle damage is noted. Superior tolerance to heat is exhibited in barley cultivars that have higher activities of antioxidant enzymes (El-Shintinawy *et al.*, 2004). In some plants, antioxidant enzyme activities are increased in response to heat shock (Edreva *et al.*, 1998; Lurie *et al.*, 1997; Tsang *et al.*, 1991). In animals, younger animals are better able to accommodate heat stress than older animals, presumably because antioxidant capacity diminishes with age (Ando *et al.*, 1997). (The heat shock proteins induced by sudden temperature changes in all organisms are not directly associated with oxidative stress, but are chaperonin proteins that assist with macromolecular folding and organization at extreme temperatures.)

Low temperature, especially when accompanied by high illumination, induces ROS production in plants (Kendall and McKersie, 1989). The enzymatic stages of photosynthesis are slowed by the low temperature, leading to excess electron excitation when light levels are high. Enhanced activity and/or levels of antioxidant enzymes thus confer tolerance to low temperature conditions. CAT prevents oxidative damage in the mitochondria of maize during low temperature stress (Prasad *et al.*, 1994) and low temperature tolerance has been increased in mutant tobacco that over-expresses SOD (McKersi *et al.*, 1993). Higher APX, CAT and radical scavenging activity conferred higher chilling tolerance in 20mm-long, relatively to 70mm-long, cucumber radicles (Kang and Sultveit, 2002). In animals, cold tolerance is often achieved by slowing metabolism even to the state of hibernation. Oxidative stress is a consequence not of cold, but of the rapid increase in oxidative metabolism when temperatures return to normal. Cold tolerance in animals uses similar pathways to ischemia tolerance (Storey and Storey, 1996).

2.8 Pollutants and Toxic Chemicals

Ozone is a phytotoxic air pollutant that is formed by the interaction of nitrogen oxides, hydrocarbons and UV radiation. Ozone has rapidly increased in level over the past few decades. Ozone is a strong oxidant that generates ROS such as H_2O_2 , O_2^-, OH and $^1O_2$ and hydroperoxyl radicals (Grimes *et al.*, 1983) once taken up by the plant through leaf stomata. Plants raised under ambient levels of ozone exhibit significant affects of oxidative stress, when compared to plants raised under ozone-free conditions (Calatayud *et al.*, 2001; 2004). Tolerance to ozone differs between species and cultivars (Pasqualini *et al.*, 2001) and depends on the ontogenetic stage of the plant (Heath, 1994). In comparative studies of ozone resistant to susceptible species it has been noted that resistance to ozone was associated with higher levels of the antioxidant enzymes including APX, POX and GR (Calatayud *et al.*, 2004; Scebba *et al.*, 2003; Pasqualini *et al.*, 2001). Ozone has both acute and chronic effects on the lung, stimulating inflammatory responses in addition to causing direct oxidative damage to lung tissues. Developing lungs are particularly susceptible to ozone-induced damage, and elevated levels of ozone and other environmental pollutants may be major contributors to the increased incidence of childhood asthma (Finkelstein and Johnston, 2004).

Anthropogenic activities such as mining, combustion of fossil fuels (including heavy vehicular traffic), application of phosphates and sewage for agricultural production, and industrial manufacturing and production lead to the accumulation of heavy metals in ecosystems. Of the 17 bio-active heavy metals (Weast, 1984), Fe, Mo and Mn are essential plant micro-nutrients, Zn, Ni, Cu, V, Co, W, and Cr are toxic but can be important trace elements and Ag, Hg, Ag, Sb, Cd, Pb, and U are more or less toxic to plants (Schützendübel and Polle, 2002). Heavy metal accumulation generally has drastic affects on shoot and root growth of economically important crop species, with likely mechanisms for toxicity including (1) induction of ROS by autoxidation and Fenton reactions; (2) blocking of essential functional groups; and (3) displacement of metal ions from essential proteins. Responses of plants to heavy metals are highly variable and depended on the plant species and tissue analysed, the metal used and the intensity of the stress. It frequently appears that antioxidant enzymes are not induced by heavy metal treatments (Gallego *et al.*, 1996), but ROS production, inhibition of respiration and ATP depletion were observed in cultured tobacco cells and pea roots treated with Al (Yamamoto *et al.*, 2002). Heavy metals can have synergistic effects (Guo *et al.*, 2004). Heavy metal toxicity is well documented in animals. There is some evidence that certain heavy metals stimulate inflammatory pathways such as the NF-κB cascade

(Chen and Shi, 2002), resulting in oxidative damage and chronic diseases such as cancer.

2.9 Nutrient Stress

Nutritional requirements of plants and animals are fundamentally different, since plants are autotrophs and animals are heterotrophs. Atmospheric CO_2 is essential for plant growth, and human activities are likely to cause a doubling in atmospheric CO_2 concentration over the next century (IPCC, 2001). As a consequence, mean global temperature is predicted to rise, rainfall patterns will change and large areas will have reduced soil moisture status. Elevated CO_2 concentration increases photosynthetic rate and water use efficiency in C3 plants, since C3 plants are often CO_2 limited due to closed stomata for water conservation. Several species including conifers and deciduous trees had lower antioxidant enzyme activities when grown under elevated CO_2 conditions, suggesting that increasing atmospheric carbon dioxide might diminish oxidative stress on plants (Polle *et al.*, 1993; 1997; Schwanz *et al.*, 1996; Marabottini *et al.*, 2001). However, the other climactic changes including drought and elevated temperatures expected to accompany changes in global CO_2 will contribute to increased oxidative stress.

Plants depend on soil and water for essential nutrients including nitrogen, phosphorus and sulfur. Inadequate soil nutrient-availability leads to plant nutrient deficit, inevitably leading to limitation of photosynthesis. This can cause over-reduction of electron transport even under conditions of low light intensity (Grossman and Takahashi, 2001). Either elevated activity of antioxidant enzymes or elevated levels of antioxidants such as ascorbic acid have been noted in plants grown under nutrient stress conditions (Tewari *et al.*, 2004; Kandlbinder *et al.*, 2004).

Nutrient stress in animals can include specific deficiencies in required nutrients such as essential amino acids, vitamins, and trace metals. These deficiencies can lead to oxidative stress especially if the antioxidant vitamins are lacking. Trace metals are essential to the activities of enzymes such as superoxide dismutase (Mn, Cu and/or Zn);

catalase (heme iron); and glutathione peroxidase (Se), but deficiencies resulting in oxidative damage are rare. Selenium deficiency occurs in China when very low selenium intake is accompanied by a viral infection (Keshan disease). In contrast to specific nutrient deficiencies, caloric restriction in animals extends life span at least in part by minimizing age-related accumulation of oxidative damage (Finkel and Holbrook, 2000).

Water is essential to all life, but plants are generally more susceptible to water-related stresses than are animals, since plants must rely on water in the surrounding environment while animals can travel to find water. In order to conserve water, plants close their stomata during conditions of drought. However this can limit CO_2 uptake necessary for photosynthesis. Consequently, the photosynthetic electron chain becomes over-reduced and, unless antioxidant capacity can control the production of ROS, lipid peroxidation occurs (Reddy, 2004). In several drought-resistant species enzyme and small molecule antioxidants increase when the plants are subjected to drought stress (Bartoli *et al.*, 1999; Zhang and Kinkham, 1996; Munné-Bosch *et al.*, 2001; Liang *et al.*, 2003) although the responses are dependent on growth conditions, stress period and plant age (Sgherri *et al.*, 2000).

Although the underlying mechanisms of salt tolerance in plants are not well understood, the capacity to cope with oxidative stress correlates with salt-tolerance in studies of rice (Dionisio-Sese and Tobita, 1998) and wheat (Meneguzzo *et al.*, 1999) varieties. In human studies there is a weak relationship between high levels of dietary salt (NaCl) and elevated blood pressure apparently due to endothelial dysfunction. Animal studies suggest that salt-induced hypertension is a consequence of increased production of superoxide, which reacts rapidly with NO making it unavailable as a vasoregulator (Dobrian *et al.*, 2003).

2.10 Anoxia

Oxygen deprivation inhibits respiration in both plants and animals, with significant consequences for overall cellular function. In plants, waterlogging can cause anoxia especially in roots. Waterlogging limits photosynthetic capacity through altered water relations, closed stomata and ultimately CO_2 limitation (Vartapetian *et al.*, 1997). Less severe ROS-induced lipid membrane damage is with anoxia-tolerance (Yordanova *et al.*, 2004; Blokhina *et al.*, 2000). High levels of antioxidant enzyme and/or metabolites are critical for survival of a number of cultivated species under waterlogging studied thus far (Lin *et al.*, 2004). In animals, ischemia—inadequate oxygen supply to a specific organ or tissue—can be a consequence of atherosclerosis or blood clots (blockage of the arteries); hypotension (low blood pressure); or a tumor or injury (compression of blood vessels). Oxidative injury accompanies reperfusion of the tissue when the tissue is resupplied with blood, and supplementation with antioxidants can limit the ischemic-reperfusion damage (Ramires and Ji, 2001).

2.11 Disease

Oxidative stress plays complex roles in disease in both plants and animals. In animals, accumulation of endogenous oxidative stress products contributes to chronic diseases including cardiovascular disease, neurodegenerative diseases such as Alzheimer's disease and Parkinson's disease, and a wide range of cancers. Although oxidative stress is not the ultimate cause of any of these diseases, accumulation of oxidative stress products contributes to their development and symptoms. In animals and plants, attack by pathogens and parasites can result in production of ROS as part of the defensive response against the pathogen (Cao *et al.*, 2001). For example, the hypersensitive response, a highly specific plant-pathogen interaction, includes an oxidative burst. The ROS that are produced may signal subsequent events in the hypersensitive response and/or may kill the infected cells and pathogens directly (Greenberg, 1997). Similarly, the immune system of animals depends on ROS to serve both as direct killers of pathogens, and as signals for further inflammatory activities. Uncontrolled, the inflammatory response is responsible for diseases such as arthritis and Crohn's disease, with destruction of healthy tissue by the feedback loop between ROS production and stimulation of the inflammation response.

2.12 Summary

The science of oxidative stress in animals is reaching maturity, but our understanding of plants remains relatively poorly developed. It is clear that there are many similarities in the oxidative chemistry of plants and animals. For example, both animals and plants are better able to tolerate environmental stresses when they are well-defended by antioxidants. We believe that understanding how environmental and genetic factors influence oxidative reactions will enable us to design crop plants better equipped to tolerate environmental stresses. In addition to permitting increased yield of crops, these antioxidant-enriched plants may have enhanced nutritive value for the consumer. Better understanding of the biochemistry behind oxidative stress responses in plants will allow us to achieve more resistant, and more nutritive, crops in the future.

References

1. Allen, D.J. *et al., J. Exper. Bot.* **49** (1998), 1775-1788.
2. Allen, R.D. *Plant Physiol.* **107** (1995), 1049-1054.
3. Ando, M. *et al., Environ. Health Perspect.* **105** (1997), 726-733.
4. Asada, K. *Physiol. Plant.* **85** (1992), 235-241.
5. Asada, K. *Causes of Photooxidative stress and amelioration of defense systems in plants.* ed. Foyer, C.H. and Mullineaux, P.M. (Boca Raton; CRC Press, 1994), 77-104.
6. Asada K. *Ann. Rev. Plant. Physiol. Mol. Biol.* **50** (1999), 601-639.
7. Baek, K-H and Skinner, D.Z. *Plant Sci.* **165** (2003), 1221-1227.
8. Bartoli, C.G. *et al., J. Exper. Bot.* **50** (1999), 375-383.
9. Blokhina, O.B. *et al., Physiol. Plant.* **109** (2000), 396-403.
10. Blokhina, O.B. *et al., Ann. Botany* **91** (2002), 179-194.
11. Bohnert, H.J. and Jensen, R.G. *Aust. J. Plant Physiol.* **23** (1996), 661-667.
12. Bor, M. *et al., Plant Sci.* **164** (2003), 77-84.
13. Bowler, C. *et al.,* Ann Rev Plant Physiol Mol Biol. **43** (1992), 83-116.
14. Calatayud, A. *et al., Photosynthetica* **39** (2001), 507-513.
15. Calatayud, A. *et al., Physiol. Plant.* **116** (2002), 308-316.
16. Calatayud, A. *et al., Photosynthetica.* **42** (2004), 23-29.
17. Cao, H. *et al., Annua. Rev.Phytopathol.* **39** (2001), 259-284.
18. Chen, F. and Shi, X. *Environ. Health Perspect.* **110** (2002), 807-811.

19. Dionisio-Sese, M.L. and Tobita, S. *Plant Sci.* **135** (1998), 1-9.
20. Dobrian, A.D. *et al., Am. J. Physiol Renal Physiol.* **285** (2003), F619-F628.
21. Doulis, A.G. *et al., Plant Physiol.* **114** (1997), 1031-1037.
22. Edreva, A. *et al., Biologia Plantarum.* **41** (1998), 185-191.
23. El-Shintinawy, F. *et al., Photosynthetica.* **42** (2004), 15-21.
24. Finkel, T. and Holbrook, N.J. *Nature.* **408** (2000), 239-247.
25. Finkelstein, J.N. and Johnston, C.J. *Pediatrics.* **113** (2004), 1092-1096.
26. Foyer, C.H. *Antioxidants in higher plants.* ed. Alscher, R.G. and Hess, J.L (Boca Raton: CRC Press, 1993), 31-58.
27. Foyer, C.H. *et al., Physiol. Plant.* **92** (1994), 696-717.
28. Foyer, C.H. *et al., Plant Cell Environ.* **17** (1994), 507-523.
29. Foyer, C.H. *et al., Plant Physiol. Biochem.* **40** (2002), 659-668.
30. Gallego, S. *et al., Plant Sci.* **121** (1996), 151-159.
31. García-Plazaola, J.I. and Becerril, J.M. *Trees* **14** (2000), 339-343.
32. García-Plazaola, J.I. *et al., New Phytologist.* **163** (2004), 87-97.
33. Gong, M. *et al., J.Plant Physiol.* **158** (2001), 1125-1130.
34. Grace, S.C. and Logan, B.A. *Plant Physiol.* **112** (1996), 1631-1640.
35. Greenberg, J.T. *Rev. Plant Physiol. Plant Mol. Biol.* 48 (1997), 525-545.
36. Grimes, H.D. *et al., Plant Physiol.* **72** (1983), 1016-1020.
37. Grisham, M.B. Reactive metabolites of oxygen and nitrogen in biology and medicine. (Austin, R.G. Lanes Co, 1992).
38. Grossman, A. and Takahashi, H. *Ann. Rev. Plant. Physiol. Mol. Biol.* **52** (2001), 163-210.
39. Guo, T. *et al., Plant and Soil* **258** (2004), 241-248.
40. Harman, D. *J. Gerontol.* **11** (1956), 298-208.
41. Harman, D. *Proceed. Natl. Acad. Sci.* **78** (1981), 7124-8.
42. Hernández, J.A. *et al., Physiol. Plant.* **89** (1993), 103-110.
43. Hernández, J.A. *et al., Plant Sci.* **105** (1995), 151-167.
44. Hernández, J.A. *et al., Plant Cell Environ.* **23** (2000), 853-862.
45. Hess, J.L. *Antioxidants in higher plants.* ed. Alscher, R.G. and Hess, J.L. (Boca Raton: CRC Press, 1993), 112-133.
46. Hull, M.R. *et al.,* Aust. J. Plant Physiol. **24** (1997), 337-343.
47. Idso, S.B. *et al.,* Agricult. Ecossys. Environ. **90** (2002), 1-7.
48. IPCC Third Assessment Report: *Climate Change 2001.* eds. Watson, R.T. and the Core Writing Team (IPCC, Geneva, Switzerland), (2001), 184.
49. Janiszowska, W. and Pennock, J.F. *Vitam. Horm.* **34** (1976), 77-105.
50. Jordan, B.R. *Adv. Botanical Res.* **22** (1996), 97-162.
51. Jung, S. *Plant Sci.* **166** (2004), 459-466.
52. Kandlbinder, A. *et al., Physiol. Plant.* **120** (2004), 63-73.
53. Kang, H.M. and Salveit, M.E. *Physiol. Plant.* **113** (2001), 548-556.
54. Kang, H.M. and Salveit, M.E. *Physiol. Plant.* **115** (2002), 244-250.
55. Kendall, E.J. and McKersie, B.B. *Physiol. Plant.* **76** (1989), 86-94.

56. Landry, L.G. *et al., Plant Physiol.* **109** (1995), 1159-1166.
57. Larson, R.A. *Phytochemistry* **27** (1988), 969-978.
58. Liang, Y. *et al., Plant and Soil.* **257** (2003), 407-416.
59. Lima, A.L.S. *et al., Environ. Exper. Bot.* **47** (2002), 239-247.
60. Lin K.H. *et al., Plant Sci.* **167** (2004), 355-365.
61. Lin, J.S. and Wang, G.X. *Plant Sci.* **163** (2002), 627-637.
62. Lurie, S. *et al., Physiol. Mol Plant Pathol.* **50** (1997), 141-149.
63. Ma, F. and Cheng, L. *Plant Sci.* **165** (2003), 819-827.
64. Ma, F. and Cheng, L. *Plant Sci.* **166** (2004), 1479-1486.
65. Marabottini, R. *et al., Environ. Pollut.* **113** (2001), 413-423.
66. McKee, I.F. *New Phytologist* **137** (1997), 275-284.
67. McKersie, B.D. *et al., Plant Physiol.* **103** (1993), 1155-1163.
68. Meneguzzo, S. *et al., J. Plant Physiol.* **155** (1999), 274-280.
69. Mittler, R. *Trends Plant Sci.* **7** (2002), 405-410.
70. Mittova, V. *et al., Free Radicals Res.* **36** (2000a), 195-202.
71. Mittova, V. *et al., Physiol. Plant.* **115** (2000b), 393-400.
72. Munné-Bosch, S. *et al., Aust. J. Plant Physiol.* **28** (2001), 315-321.
73. Munné-Bosch, S. *Plant Sci.* **166** (2004), 1105-1110.
74. Noctor, G. and Foyer, C.H. *Ann. Rev. Plant. Physiol. Mol. Biol.* **49** (1998), 249-279.
75. Pallett, K.E. and Young, A.J. *Antioxidants in higher plants.* ed. Alscher, R.G., Hess, J.L. (Boca Raton: CRC Press, 1993), 31-58.
76. Paolacci, A.R. *et al., J.Plant Physiol.* **150** (1997), 381-387.
77. Pasqualini, S. *et al., Plant Cell Environ.* **24** (2001), 245-252.
78. Pereira, G.J.G. *et al., Plant and Soil* **239** (2002), 123-132.
79. Polle, A. *et al., Plant, Cell Environ.* **16** (1993), 477-484.
80. Polle, A. and Morawe, B. *Botanica Acta* **108** (1995), 314-320.
81. Polle, A. *Plant Physiol.* **126** (2001), 445-462.
82. Prasad, T.K. *et al., Plant Cell.* **6** (1994), 65-74.
83. Purvis, A.C. *Physiol. Plant.* **100** (1997), 165-170.
84. Ramires, P. and Ji, L.L. *Am. J. Physiol.* **281** (2001), H679-H688.
85. Reddy, A.R. *et al., Environ. Exper. Bot.* **52** (2004), 33-42.
86. Scebba, F. *et al., Environ. Pollution* **123** (2003), 209-216.
87. Scebba, F. *et al., Plant Sci.* **165** (2003), 583-593.
88. Schützendübel, A. and Polle, A. *J. Exper. Bot.* **53** (2002), 1351-1365.
89. Schwanz, P. *et al., Plant Physiol.* **110** (1996), 393-402.
90. Sgherri, C.L.M. *et al., J. Plant Physiol.* **157** (2000), 273-279.
91. Storey, K.B. and Storey, J.M. *Annu. Rev. Ecol. Syst.* **27** (1996), 365-386.
92. Streb, P. *et al., J. Exper. Bot.* **54** (2002), 405-418.
93. Strid, A. *et al., Photosynthetic Res.* **39** (1994), 475-489.
94. Tewari, R.K. *et al., Plant Sci.* **166** (2004), 687-694.

95. Thomas, C.E. *et al.*, *Lipids* **27** (1992), 543-550.
96. Tolbert, N.E. *Annu. Rev. Biochem.* **50** (1981), 133-157.
97. Tsang, E.W.T. *et al.*, *Plant Cell* **3** (1991), 783-792.
98. Vaidyanathan, H. *et al.*, *Plant Sci.* **165** (2003), 1411-1418.
99. Vartapetian, B.B. and Jackson, M.B. *Annals Bot.* **79** (1997), 3-20.
100. Verdaguer, D. *et al.*, *Plant Cell Environ.* **26** (2003), 1407-1417.
101. Vranová, E. *et al.*, *J. Exper. Bot.* **53** (2002), 1227-1236.
102. Weast, R.C. *CRC Handbook of chemistry and physics.* (64th ed Boca Raton CRC Press, 1984).
103. Wise, R.R. *Photosynthesis Research* **45** (1995), 79-97.
104. Yamamoto, Y. *et al.*, *Plant Physiol.* **128** (2002), 63-72.
105. Yan, B. *et al.*, *Plant and Soil* **179** (1996), 261-268.
106. Yordanova, R.Y. *Environ. Exper. Bot.* **51** (2004), 93-101.
107. Zhang, J. and Kirkham, M.B. *New Phytologist* **132** (1996), 361-373.

CHAPTER 3

THE EXERCISE CONTINUUM

Ron L. Wiley
Miami University, Oxford, OH

3.1 Introduction

Physical activity is an umbrella term that includes movement by the contraction of skeletal muscle. Leisure activities including exercise and sports are forms of physical activity. It is indisputable that health can be compromised, maintained, or improved depending on the level of exercise or physical activity that is (or is not) performed on a regular basis. Most people, including scientists, when asked the question "What is exercise?" will respond with an answer that includes or implies movement. Some will explain or define exercise by creating a list of activities, such as running, jumping, walking, cycling, rowing, swimming, and the like. The most popular type of exercise is aerobic exercise, which usually involves dynamic, isotonic muscle movements. Other modes of exercise involve different types of muscle contractions. Isometric exercise is one type that occurs without muscle fiber movement as when pushing against an immovable object or when squeezing a handgrip device (Fig. 3.1).

Fig. 3.1. Handgrip exercise is an example of isometric muscle contraction.

All types of exercises, whether they involve movement or not, confer health benefits (Wiley *et al.,* 1993). Isometric and isotonic contractions are at different ends of an exercise continuum. While much force is generated during isometric exercise, work cannot be quantified in the traditional way that isotonic exercise is because the muscles do not move a measurable distance. So the usual formula [Eq.(3.1)]:

$$Work = Force \ X \ Distance \qquad (3.1)$$

cannot apply to isometric contractions. Instead the equation [Eq.(3.2)]:

$$Work = Force \ X \ Duration \ of \ Contraction \qquad (3.2)$$

more accurately applies to isometric muscle contractions.

Isometric exercise efforts are easily quantified. With encouragement, subjects can give very reliable and repeatable maximum voluntary contractions (MVC) as the "yardstick" against which all efforts can be measured. Although some studies averaged the resulting forces after subjects performed two or three maximum efforts with rest periods between, the single greatest force should be used as the maximum. By recording the MVC, targets that are a percentage of MVC can be set. Some may exert a contraction = 50% of MVC while others exert a contraction = 75% of MVC, depending on capability and goals. Subjects can then hold this isometric contraction for either prescribed durations or to the point of fatigue. The fatigue point is usually defined as the point at which the sustained force drops below some pre-determined level, such as 10% below the original target. In experiments in which targets will be set on succeeding days, a new MVC is determined each day because the muscle group being exercised will show an increase in strength.

Few studies have measured the strength increases that can be achieved by isometric exercise. Komi *et al.* (1978) reported an average increase in isometric knee extension force of 20% with isometric training. Several reports of increased isometric strength in older individuals who participated in *resistance* training are summarized by

Rogers and Evans (1993), Improvements in strength ranged from 9 – 174% of their original strength measurements.

It is accurate to state that most muscle movements are neither totally isometric nor totally isotonic, but some combination of both. Here, a concept is proposed that there is a continuum of exercise, with isometric efforts anchoring one end and isotonic or dynamic efforts anchoring the other (Fig. 3.2). Varying proportions of isometric and dynamic muscle contractions are shown in the middle of the continuum as the weight lifter raises the bar overhead. During the lift, muscle contracts dynamically and at the end of the lift when the weight is held overhead, the muscles exert force without movement, which is an isometric contraction.

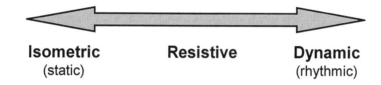

Isometric   **Resistive**   **Dynamic**
(static)   (rhythmic)

Fig. 3.2. Exercise continuum with isometric or static exercise on one end and dynamic or rhythmic exercise on the other end.

3.2 Isometric Exercise

The literal definition of "equal length" implies that muscle does not change length during a contraction. When muscles generate force without movement or a change in joint angle that is considered an isometric contraction. While true for the overall length of the muscle from origin to insertion, there must be some shortening of muscle fibers to take up the slack in series and parallel elastic components and place force on the tendons and bones involved. In an isometric contraction actin and myosin filaments attach and form cross-bridges, but there is relatively little sliding of actin alongside myosin and sarcomere shortening as slack is taken up in elastic elements and the isometric force is generated. Sustained isometric efforts include squeezing a hand grip, pushing against a wall or other immovable object; carrying a heavy object with the hands/arms; or sustaining a large mass overhead. As the velocity of muscle movement approaches zero, more force can be generated during concentric or muscle shortening contractions. The opposite relation exists in eccentric or muscle lengthening contractions where more force is generated when velocity of contractions increase (Fig. 3.3).

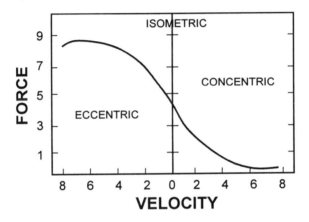

Fig. 3.3. Force-velocity relationships for isometric and two types of isotonic muscle contractions: eccentric (lengthening) and concentric (shortening).

3.3 Isotonic Exercise

Consider the literal definition of "equal tone" (or force). This implies that there is movement, but that a nearly constant tone, or force, is applied throughout the range of motion. While this is not exactly true, it is nearly so during many efforts such as throwing. But there certainly is variation in tone and force production throughout the entire range of motion. Two different types of isotonic muscle contractions are concentric and eccentric.

Concentric muscle contraction is a form of isotonic exercise. It involves muscle shortening whereby the actin filaments are pulled closer together by myosin filaments, pulling the z-lines closer towards the middle of the sarcomere. (Fig. 3.4) Peak force production during concentric muscle contractions can be enhanced if the muscle is stretched 20% before initiating movement and when contractions occur slowly. The slower a concentric contraction, the greater the force production.

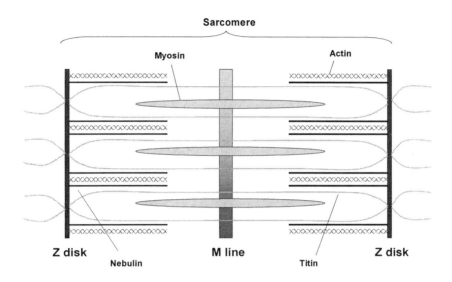

Fig. 3.4. Schematic diagram of sarcomere with contractile proteins actin and myosin, and architectural proteins nebulin, and titin.

Eccentric exercise is also a form of isotonic exercise. This form of isotonic exercise occurs passively during the re-setting of concentric muscle shortening. An example is when a force is being lowered as occurs in an arm extension. In this case, no major physiological challenge occurs to the muscle which is contracting at submaximal force. When a peak contraction effort is attempted during eccentric contraction, the muscle lengthens and the actin filaments make contact with the myosin filaments, but pull further away from the center of the sarcomere. Compared with muscle shortening, more peak force can be generated when the muscle contracts by lengthening. Furthermore, contrary to concentric contractions, velocity is directly related to force production during eccentric muscle contractions. So, the faster the eccentric movement, the greater the force production.

In the exercise continuum model depicting two extreme types of exercise, we also see distinct oxygen demands, energy transfer, and potential for oxidative stress (Fig. 3.5). During isotonic exercise, rhythmic muscle contractions usually last more than 10 seconds and require oxygen in the metabolism of lipids and glycogen or glucose for energy transfer.

Isometric exercise does not rely on elevated oxygen uptake for energy transfer. Instead the force generated via isometric exercise uses short term energy from the phosphagen system: adenosine triphosphate (ATP) diphosphate and creatine phosphate (CP). ATP stores will break down to adenosine (ADP) and supply energy for only 1 to 2 seconds. CP stores will replenish ATP by donating its phosphate (P) to ADP to form ATP in the presence of an enzyme creatine phosphokinase. This process will continue for approximately 8 to 10 seconds. This 10 second exercise duration usually includes high intensity work that results in high ADP and energy transfer. The high ADP levels can act as substrate for ROS just as high oxygen uptake levels do in isotonic exercise, via hypoxanthine (HX) and xanthine (X) that are produced from ADP.

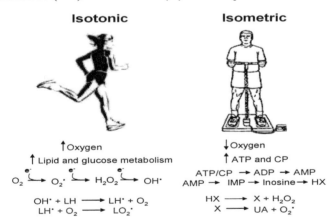

Fig. 3.5. Isometric and isotonic exercise have different oxygen demands, primary energy substrates and oxidative stress reactions $O_2 \cdot$: superoxide radical; OH•: hydroxyl radical, LH= lipid molecule, LH•: lipid radical; $LO_2 \cdot$: lipid peroxide radical; ATP: adenosine triphosphate; ADP: adenosine diphosphate; AMP: adensoine monophosphate; IMP: inosine monophosphate; HX: hypoxanthine; X: xanthine; H_2O_2: hydrogen peroxide; UA: uric acid.

3.4 The Exercise Continuum

One can understand the concept of the exercise continuum, by first asking the question: what are the extremes? For isometric exercise, it's rather easy to provide examples; anything in which a force is applied with no movement and, therefore no external work is performed. This readily satisfies the physicist, but the physiologist will quickly remind us that metabolic work is certainly done. Examples have already been given, including a sustained handgrip squeeze.

As long as gravity is present, there will be an isometric component in virtually all muscles in the body. Therefore, consider an astronaut in microgravity moving limbs to accelerate the body. This would be a purely isotonic exercise. Consider pushing off a wall or object to accelerate through the capsule or through space during an extravehicular activity. The isometric component at the time of the push would be quite small, and the limb movement almost purely rhythmic (Fig. 3.6).

Now let's develop the large middle portion of the continuum line. Imagine pedaling a 10-gear bicycle up a gradient. First, start at the bottom of the hill and pedal up in first gear (Fig. 3.7). The rhythmic component is large, with the rapid pedaling, and the isometric component, which is the instantaneous push downward on the pedals, is relatively small. Now imagine starting at the bottom of the hill in 10th gear (Fig. 3.8). The isometric component (push on the pedal) is now large and the rhythmic one (rapid pedal cycling) is small. Other activities fit on the continuum depending upon the relative components. Walking or running has a smaller isometric component – that of supporting the mass of the body against gravity. Lifting two-pound free weights with the arms would consist of a larger rhythmic component and a smaller isometric component, while pressing one's maximum weight overhead would have the reverse.

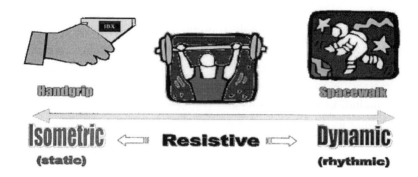

Fig. 3.6. Exercise continuum with hand squeeze at "pure" isometric end and astronaut moving limbs at "pure" rhythmic end.

Fig. 3.7. Easy effort simulating dynamic exercise.

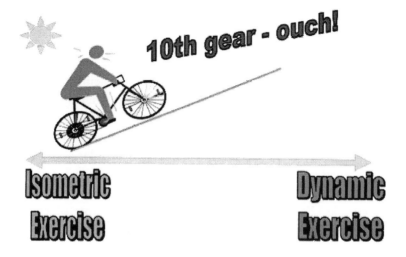

Fig. 3.8. Difficult effort simulating isometric exercise.

Fig. 3.9. Curling free weights is a more rhythmic or isotonic contraction compared with pressing a weight overhead, which is a more isometric contraction.

Another "category" term that has come into use is "resistive" exercise training. On the continuum line this would simply be placed closer to the isometric end, as it implies a load that gives relatively high resistance to movement. Accepting the muscle continuum concept for muscle contractions allows descriptions of the physiological responses to exercise to be interpreted based upon the recognition of the proportional components of isometric/isotonic combinations. For example, the greater the isotonic component, the greater the expected increase in oxygen uptake. The greater the isometric component, the greater the increase in ATP breakdown and ischemic-reperfusion response. These actions affect mechanical and metabolic stress differently. Nevertheless, along the exercise continuum, it is well accepted that acute exercise, whether it be mostly isometric or rhythmic, increases oxidative stress. Chronic exercise, that is, regular performance of either isometric or rhythmic types of exercise, results in adaptive responses that protect muscle from oxidative stress reactions. Over a lifetime, muscle changes in shape and function. The role of exercise across the continuum from isometric to isotonic, on oxidative stress and aging will be explored in the following chapters.

3.5 Summary

Along the exercise continuum, it is well accepted that acute exercise, whether it be mostly isometric or rhythmic, increases oxidative stress. Chronic exercise, that is, regular performance of either isometric or rhythmic types of exercise, results in adaptive responses that protect muscle from oxidative stress reactions. Over a lifetime, muscle changes in shape and function. The role of exercise across the continuum from isometric to isotonic, on oxidative stress and aging will be explored in the following chapters.

References

1. Wiley, R.L. *et al.*, *Med. Sci. Sports Exerc.* **24** (1992), 749-754.
2. Komi, P.V. *et al.*, *Eur. J. Appl. Physiol. Occup. Physiol.* **40** (1978), 45-55.
3. Rogers, M.A. and Evans, W.J. *Exercise and Sports Science Reviews.* ed. Holloszy, J.O. **21**(1993), 90-95.

CHAPTER 4

OXIDATIVE STRESS AND MUSCLE SIZE, TYPE, AND ACTION

H.M. Alessio
Miami University, Oxford, OH

4.1 Introduction

Muscles, their size, type, and functional actions, are critical to the lives of higher animals. The ability to store and metabolize energy as well as to generate force, heat, and movement allows animals to evade predators, forage for food, build shelters, digest food, adjust body temperature, grow, develop, reproduce, and repair damaged cells. Throughout evolution the chances for survival increased dramatically in animals having the largest, strongest, and fastest muscles. Large muscles provide a mechanical advantage for seeking food, water, and shelter, and a metabolic advantage that can transfer large amounts of energy when needed.

Cutler (1984) compared body size with lifespan in animals living in their natural environments (Fig. 4.1). Muscle accounts for the majority of body mass and the relation between muscle mass and life span was found to be a direct one. He described an exponential shape for a mammalian survival curve where the shortest-lived animals such as shrews and mice had the least amount of muscle mass and the longest-lived animals such as the chimpanzee and human had the largest amount of muscle mass. In addition to the metabolic and mechanical advantages of large-muscled animals, survival increased in animals that were better able to adapt to new environments.

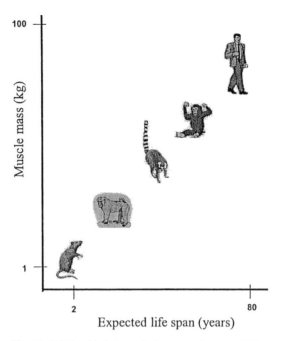

Fig. 4.1. Relationship between body mass and expected lifespan.

4.2 Oxidative Stress in Large and Small Muscles

The neuromuscular system relies on oxygen to support metabolic reactions associated with neural activation, energy transfer, growth, and repair. Skeletal muscle accounts for a moderate portion of the body's total oxygen consumption at rest (about 20%), which increases to a much larger portion during physical exercise (up to 80%). Nerves and muscles are post-mitotic tissues incapable of further cell division. The repair capacity of nerves and muscles is relatively low. This does not bode well for nerves' and muscles' vulnerability to oxidative stress, which has been formulated by Weindruch (1995) as:

Oxidative Stress = Free Radical Exposure – (Removal + Repair)

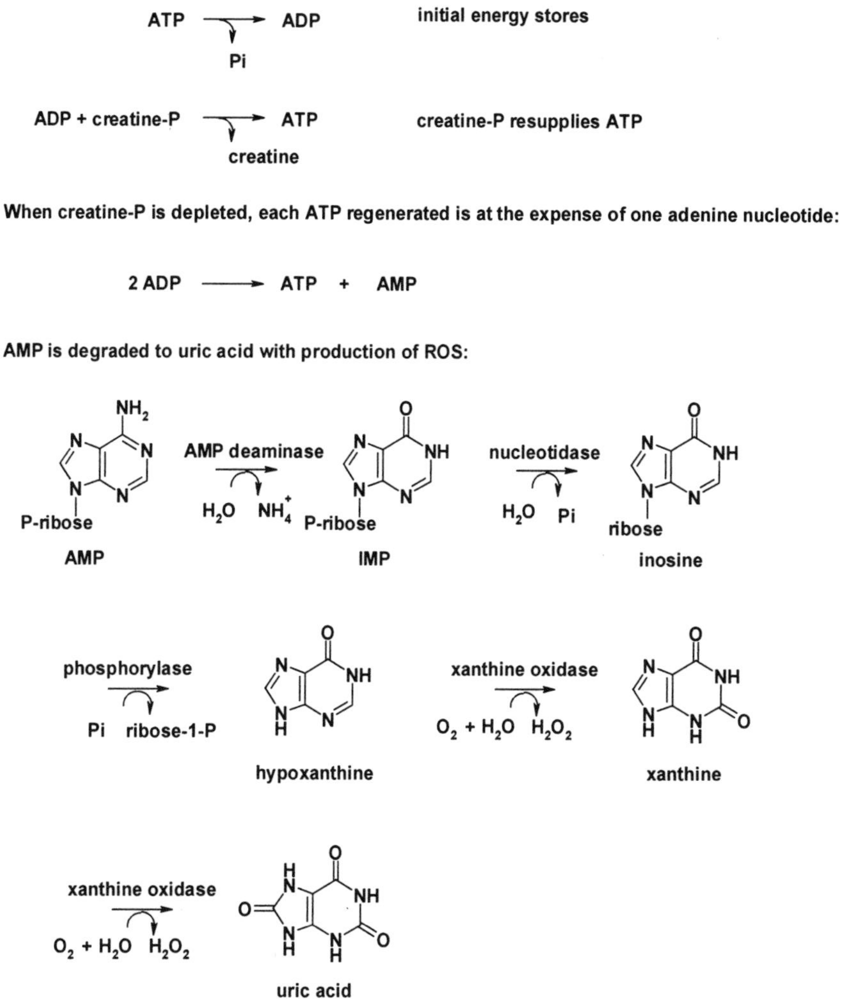

Fig. 4.2. Pathways for regeneration of ATP in muscle, and decomposition of AMP to uric acid. Xanthine oxidase either produced H_2O_2 as shown, or in the presence of NAD^+ produces $\cdot O_2-$.

Animals with large muscles can potentially generate more free radicals and other ROS compared with smaller animals. ROS may be higher in larger animals because of their greater muscle mass. Large muscles consume more oxygen than small muscles because large muscles contain more mitochondria, have a greater blood supply, and require more oxygen to support the energy demands associated with metabolism of muscles at rest and during physical activity.

Aerobic, energy-producing reactions occur in the mitochondrial electron transport chain as part of a complex process whereby oxygen is reduced by four electrons to form water. This occurs at the same time that adenosine diphosphate (ADP) is phosphorylated to adenosine triphosphate (ATP). However, the reduction of oxygen does not always occur as planned and a small fraction of electrons escape from the electron transport chain. That means that O_2 is not fully reduced to water, and some oxygen molecules are left in reactive, partially reduced states (Yu, 1996).

Superoxide, $\cdot O_2$-, is the likely initial ROS produced in mitochondria. While the amount of $\cdot O_2$- formed this way is small (1 to 5% of all oxygen that is consumed), it is nevertheless proportional to oxygen consumption (VO_2). Large animals will take in a greater absolute amount of oxygen via larger lungs, heart, and skeletal muscle. So a mass action effect is believed to exist whereby the higher the VO_2 the higher the $\cdot O_2$- produced inside the cell. Fridovich and Freeman (1986) estimated that the formation of ROS is approximately 50 nmol per gram of tissue per minute. Large animals could theoretically produce up to 100 times more ROS than a small animal. A mathematical comparison would have a 500 gram rat forming 36 mmol ROS per day and a 75 kg human forming 5400 mmol ROS per day. If endogenous antioxidants are proportional to ROS produced in the different sized animals, then large animals would not be at any greater risk than small animals for oxidative stress.

Another mechanism by which large animals may generate more ROS compared with small animals is due to greater ATP stores in their larger muscles. ATP can be degraded to $\cdot O_2$- and H_2O_2 via xanthine oxidase mediated reactions (Fig. 4.2). This is more likely to occur under ischemic-reperfusion conditions. That could include high intensity exercise where muscle contractions occur with little recovery to facilitate reperfusion to the muscle. When muscle relaxation occurs and blood flow is restored, reperfusion is often accompanied by an influx of oxygen that promotes radical formation (McBride and Kraemer, 1999).

Muscle ATP stores range from 5 to 8 mmol/kg wet weight. The Vastus lateralis muscle weighs approximately 1 gram in a rat and 2 kg in human, so a difference of approximately 2000-fold exists between rat and human ATP levels in this muscle. The higher potential for ROS production from large ATP stores could be met with proportionally higher antioxidant activity in large animals. Studies of different-sized animals across the animal kingdom have shown that the antioxidant defense systems in large animals seem to be well supplied to defend against ROS generated under both normal resting and stress-related conditions. Cutler's (1984) reports of direct correlations between antioxidants and life span in different animal species suggest that defense mechanisms are in place in bigger mammals to fend off potentially dangerous ROS formed by either higher VO_2 or ATP stores.

Antioxidants developed over the last 3 billion years to detoxify ROS and RNS (Sen and Packer, 2000). Although enzymatic and nonenzymatic antioxidant defense are found at relatively low concentrations in resting skeletal muscle levels do "adapt" to exercise-induced oxidative stress, usually by upregulating during acute exercise. Superoxide dismutase (SOD) and catalase (CAT) work together in the following reactions [Eqs. (2.1) and (2.2)]:

$$2\,O_2^- \ + \ 2\,H^+ \ \xrightarrow{\quad SOD \quad} \ H_2O_2 \ + \ O_2 \qquad (2.1)$$

$$2\,H_2O_2 \ \xrightarrow{\quad catalase \quad} \ 2\,H_2O \ + \ O_2 \qquad (2.2)$$

Production of H_2O_2 by SOD is not necessarily advantageous to the cell, since it can be converted to $OH\bullet$, but CAT dehydrates it to H_2O. Together, SOD and CAT act to reduce ROS in a fashion that appears to impact maximum life span (Orr and Sohal, 1994). SOD, CAT, and some small molecule antioxidants such as uric acid have linear or near-linear correlations with life span potential in mammals (Cutler, 1991) (Fig. 4.3).

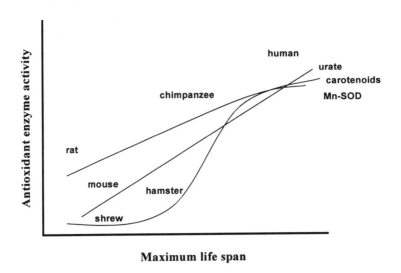

Fig. 4.3. Serum antioxidants and maximum lifespan in animals.

4.3. Metabolism and Oxidative Stress in Small and Large Animals

The abundance of antioxidants in large animals may explain the added protection they have compared with smaller animals. Paradoxically, animals with smaller muscles actually consume more oxygen per kg body mass than larger animals. Cutler described the net rate of production of free radicals in a given tissue as its specific metabolic rate. Figure 4.4 includes a few key data points from his work (1984) showing a relation between specific metabolic rate and life span potential. The larger the animal size the lower the specific metabolic rate.

This inverse relationship is explained by less heat energy being required to maintain body temperature in large animals having body surface areas that are proportionally smaller when compared with their very large body weight. A small light-weight animal will have proportionally more body surface area compared to their low body weight and will therefore have to produce more heat energy to maintain body temperature compared to a larger animal. Another measure that reflects metabolism and distinguishes small and larger animals is heart rate. A rodent's resting heart rate exceeds 400 beats per minute while a human's resting heart rate is approximately 70 beats per minute. The increased metabolic rate as reflected by specific metabolic rate and heart rate in small animals compared with larger animals implies that small animals are more vulnerable to ROS than larger animals for at least two reasons: (1) they generate ROS at a higher rate for their size and (2) they have lower levels of endogenous antioxidants to defend against ROS. These two distinctions may partly explain differences in the aging rates and life spans between small and large animals.

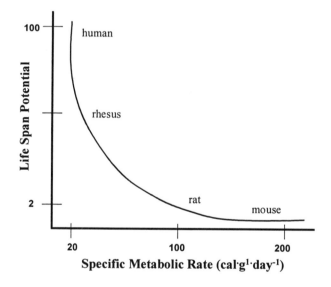

Fig. 4.4. Relation between life span potential and specific metabolic rate in four different sized animals.

Although many antioxidants have been found to increase in proportion to animal size, it should not be expected that all antioxidants are found in higher quantities in long-lived species. GSH, CAT, and plasma ascorbate are examples of antioxidants that do not show linear or near-linear correlations with life span potential in mammals. Cutler (1991) explained that simply evaluating antioxidant levels is not the only way to differentiate between short and long-lived animals. Other strategies to effectively counter oxidative stress reactions include a decreased rate of oxygen radical production which may depend on a lower VO_2, a tightly controlled electron transport chain, as well as cell and tissue traits that vigorously defend against radical attacks.

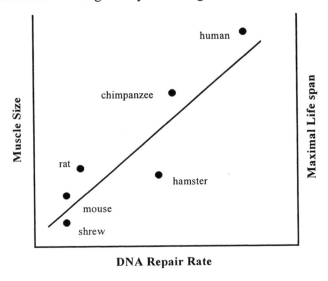

Fig. 4.5. DNA repair rate and animal size and maximal life span

4.4 Radical Damage and Damage Control in DNA

Antioxidants provide protection against radial attack on different parts of the muscle cell: the sarcolemma's phospholipid bilayer, mitocondrial membranes, nuclear membranes, sarcoplasmic reticulum, lipid, protein, and glucose substrates, and the contractile proteins: actin and myosin. This protection is not fail-safe. Sometimes muscles experience trauma, either metabolic or mechanical, that overwhelm endogenous antioxidant defenses. If damaged, muscles can undergo repair by turning on specific genes that activate protein turnover-removal of damaged fragments and synthesis of new proteins (see Chapter 9). Cutler (1984) reported that the ability to repair DNA correlated directly with animal size and life span (Fig. 4.5). This relation suggests that general or specific DNA repair mechanisms are in place in large animals to protect and stabilize genes and assure that RNA and protein synthesis proceed with as few errors as possible.

4.5 Muscle Type and Oxidative Stress

At least seven different fiber types have been identified in human skeletal muscle (Staron, 1997). More fiber types are likely to exist among different species across the animal kingdom. That is because muscle fiber type is influenced by genetics, environment, hormones, neural stimulation, food intake, physical activity, gender, and age. Skeletal fiber types are established shortly after birth. Nevertheless, over the course of a lifetime, transient muscle fiber types come and go depending on environmental conditions that affect frequency and intensity of neural-muscular activation which influence myosin heavy chain content, sarcoplasmic reticulum calcium uptake and activities of calcium ATPase, proteolytic caspase, cytochrome c-oxidase, succinate dehydrogenase, and glycolyolytic enzymes. These factors affect muscle type because they regulate ATP levels, oxygen handling capacity, substrate utilization, protein turnover, and amount of force generation in muscles.

Although seven different muscle fiber types have been identified, usually skeletal muscle is categorized into two major fiber types: (1) slow twitch oxidative (Type I) and (2) fast twitch (Type II). Type I contains higher levels of mitochondrial proteins and oxidative enzymes compared with Type II fibers. Depending on the muscle fiber type, a muscle may contain anywhere from less than a hundred to thousands of mitochondria. Mitochondria are organelles found in the trillions of cells throughout the body. They contain enzymes that break apart different substrates (glucose, fats, proteins) and release energy from reactions that occur along their inner membranes. Of the two main mitochondrial membranes, the outer and the inner compartment, it is the inner compartment that processes oxygen (Fig. 4.6).

Oxygen supply to Type I fibers is greater than Type II because of increased capillarization surrounding Type I muscle. Intramuscular oxygen transport within Type I fibers is enhanced via elevated myoglobin levels, which gives Type I fibers its signature red color. The substrate of choice in Type I fibers is fatty acids derived from relatively large triglyceride stores found in close proximity to the mitochondria. In the process of producing ATP, increased oxidative enzymes in Type I muscle act in redox reactions inside the mitochondria and reduce oxygen to water. Type I muscle fibers are the first to be recruited during exercise, are the major fiber type used during submaximal exercise, and are the last to deplete their energy stores, which consist primarily of triglycerides and glycogen. Endurance training can improve the aerobic capacity of Type I muscle fibers (Kirkendall and Garrett, 1998).

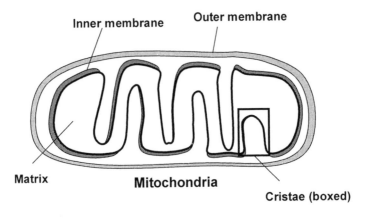

Fig. 4.6. Mitochondria-the inner membrane is where oxygen is processed.

At least two categories of fast twitch fibers have been identified: Type IIa (fast twitch oxidative) and Type IIb (fast twitch glycolytic). They differ by their ability to exert large amounts of force, fatigability, and the oxygen handling capacity. Compared with Type IIb, Type IIa muscle fibers generate less force per time, are less fatigable, and have more myoglobin, giving it its signature pink color. The additional oxygen handling capacity of Type IIa compared with IIb fibers are supported by relatively higher levels of mitochondrial proteins and oxidative enzymes. The substrate of choice for Type IIa fibers is glycogen and for Type IIb fibers is phosphagen system molecules such as ATP and CP. Type IIb fibers have a better formed sarcoplasmic reticulum that is capable of quickly releasing and taking up Ca^{2+} during fast contractions. Type II fiber activation usually follows Type I fibers, which are recruited first during submaximal work. Then Type IIa followed by Type II b fibers are recruited usually - in that order - for bursts of energy during progressively more intense exercise associated with isometric or isotonic resistance exercise, sprinting, and muscle contractions required to achieve a maximal endurance exercise effort (Fig. 4.7).

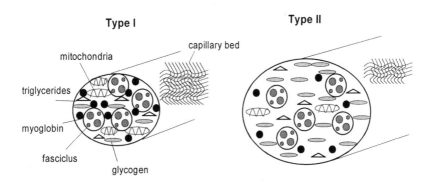

Fig. 4.7. Cross section of Type I and Type II muscle fibers showing differences in myoglobin, glycogen, mitochondria, triglycerides, and capillaries.

Natural selection accounts for the distribution of fast and slow twitch fibers across the animal kingdom and within each species. Survival for some animals depends on speed; for others, endurance; and for others, a combination of both speed and endurance. Inheriting genes that code for the "right" type of muscle fibers in animals that live in a certain environment with specific predators, food sources, and terrain will determine the chances for survival and for passing along valuable genes to the next generation. Some adaptations can occur in muscle, although the transformation from one muscle fiber type to another does not normally occur.

The average human is born with 50% slow and 50% fast twitch muscle fibers, but elite endurance athletes have up to 80% Type I muscle fibers and elite strength and power athletes have up to 80% Type II muscle fibers (Costill *et al.*, 1987). Different types of exercise training can influence the characteristics of muscle fibers and change the proportions of fiber types. For example, sprint training has been shown to decrease the proportion of Type I and increase the proportion of Type IIb muscle fibers. Endurance training will increase the proportion of Type I and increase the proportion of Type IIa relative to Type IIb. These changes in proportions of skeletal fiber types occur by metabolic shifts that include changes in glycogen and fat metabolism, myofibrillar and myosin ATPase activities, and changes in oxidative enzyme activities (Roy *et al.*, 1985).

Type I fibers appear to experience less oxidative stress compared with Type II. Soleus muscle, in particular, usually demonstrates lower biomarkers of lipid peroxidation (e.g. thiobarbituric acid reactive substances (TBARS), lipid hydroperoxides (LH), malonaldedyde (MDA)) compared with Type II fibers (Oh-Ishi *et al.*, 1995; Alessio *et al.*, 1988). Antioxidant activities of GSH peroxidase (GPX), GSSG reductase (GR), and CAT, but not SOD were significantly higher in Type I soleus compared with Type II (Ji *et al.*, 1992). Muscle Mn-SOD (found in mitochondria) as well as GPX was higher in Type I soleus compared with Type II vastus muscles (Oh-Ishi, 1995). It seems some combination of selective muscle fiber recruitment, antioxidant activity, oxidative enzyme activity, lipid peroxidation initiating or clearing factors and some other factors play a role in distinguishing Type I and Type II fiber susceptibility to ROS.

Muscle oxidative characteristics and recruitment patterns during exercise are major regulators in oxidative stress balance. GPX, SOD, and lipid peroxidation were compared in Type I and Type II muscle fibers following 60 minutes of exhaustive running (Caillaud *et al.*, 1999). Type I muscle did not show major shifts toward oxidant activity. This might be due to adequate levels of antioxidant protection to cope with exercise-induced ROS. In the more glycolytic Type II muscles, ROS production is likely to be lower (due to lower oxygen supply and mitochondria), and therefore less antioxidant activity is needed. Nevertheless high intensity exercise has been reported to increase plasma TBARS and LH in both Type I and II muscle fibers (Goldfarb *et al.*, 1994; Alessio *et al.*, 1988). Moderate intensity exercise has been reported to increase MDA in Type I (62% above resting levels) and Type II (90% above resting), however the most oxidative fiber, the soleus, experienced the least increase in MDA of all muscle fiber types (Alessio *et al.*, 1988). Most studies agree that the most oxidative muscles (e.g. soleus) have the most ROS and the most antioxidant activity for defense (Caillaud *et al.*, 1999; Lawler *et al.*, 1993; Ji *et al.*, 1993; Alessio and Goldfarb, 1988).

4.6 Muscle Remodeling and Oxidative Stress

In order for muscles to adapt to exposure to stress, a cycle of protein degradation followed by protein synthesis must occur (Fig. 4.8). Muscle remodeling occurs by proteolytic processes that remove proteins damaged by mechanical or metabolic stress which can occur by over or under-use. This is followed by specific second messenger kinases that turn on genes that express for new contractile proteins to replace and possibly add to existing and damaged skeletal muscle proteins.

In order for muscle remodeling to take place, both protein degradation and protein synthesis must occur. Degradation can occur as a consequence of either overuse or disuse. Three different proteolytic systems have been described in skeletal muscle degradation: (1) ubiquitin-proteasome, (2) lysosomal cathepsins, and (3) Ca^{2+} dependent calpains.

The ubiquitin-proteasome system has a major role in protein degradation especially as a result of disuse. Damaged proteins are identified and marked for degradation by ubiquitin-activating (E1s), ubiquitin-conjugating (E2s), and ubiquitin protein ligase (E3s) enzymes (Jagoe and Goldberg, 2001). Proteasomes are activated by these different types of ubiquitin and proceed to degrade myofibrillar proteins that are released from the sarcomere-probably because they were damaged.

The lyosomal cathepsins system is thought to have a major part in protein degradation in the late stages of muscle atrophy (Stevenson *et al.*, 2003). The activation of some cathepsins, C, D, and L, and inhibition of cathepsin B appear to work together over a period of days to break down protein.

Calpains are calcium-activated proteases that regulate protein turnover in cells. Ca^{2+} dependent calpains are believed to play a role in the breakdown of myofibrillar proteins by proteasomes (Huang & Forsberg, 1998). Different calpain subgroups (e.g. 2, 3) and their modulators (e.g. diazepam-binding inhibitor) appear to respond at different times, directions, and intensities to muscle atrophy from disuse (Stevenson *et al.*, 2003). In general Ca^{2+} activated calpains increase when muscle is altered from disuse. Calpains are also implicated in apoptotic cell death (Moore *et al.*, 2002), but the mechanism or mechanisms by which calpains act are complex.

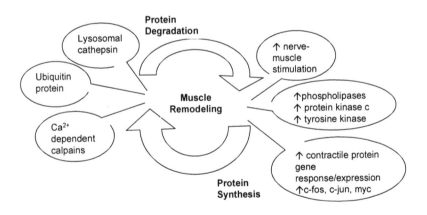

Fig. 4.8. Factors contributing to muscle remodeling.

Protein degradation and removal by any of these three pathways is an essential part of muscle remodeling. By-products of muscle breakdown act as a stimulus for protein synthesis. Muscle growth occurs mainly by increased fiber size and in some cases, increased fiber number. The larger fiber size is due to an increase in the amount of contractile proteins, namely actin and myosin. These proteins are regulated by a family of genes that include c-fos, c-jun, and myc. The second messengers: phospolipases, protein kinase C, and tyrosine kinase, convey increased neural-muscular stimulation to the genes that ultimately express for actin and myosin.

4.7 Oxidative Stress and Muscle Action

During muscle contractions, actin and myosin generate static force or are pulled and pushed, generating tension alongside structural proteins (e.g. titin, nebulin, alpha-actininen) inside the sarcomere. The amount of force generated during muscle contraction, whether isometric or isotonic, temporarily changes the shape of muscle fibers. During muscle contraction, a series of steps occur that change virtually all components of the internal environment inside the sarcomere. Some of these changes include increased release of radical molecules in the motor neuron and the muscle fiber.

In order for a muscle to contract, first, an electrical impulse must travel through a motor nerve to the axon terminal. Axon terminals are located at the end of the nerve and close to the sarcomere. The sarcolemma physically separates the nerve from the muscle and contains receptors that are sensitive to over 50 neurotransmitters. Acetylcholine (Ach) and norepinephrine (NE) are two primary neurotransmitters that transmit nerve impulses from the axon terminal to the sarcolemma to initiate muscle contractions.

When an electrical impulse arrives at the axon terminal, the nerve ending secretes Ach into the space between the nerve and muscle. Ach binds to specific receptors on the sarcolemma. Until Ach-receptor binding exceeds a certain activation level, the sarcolemma maintains a resting membrane potential of -70 mV. The enzyme Na^+-K^+ ATPase (also referred to as the sodium-potassium pump) maintains this resting membrane potential by selectively transporting 3 Na^+ out of the cell for every 2 K^+ it transports into the cell.

Changes in the membrane potential occur when the sodium pump changes and Na$^+$ and K$^+$ exchange differs from the usual 3 Na$^+$ out, 2 K$^+$ in. This depolarizes the sarcolemma, changing the membrane potential from -70 mV towards 0. The change from electronegativity to electropositivity will open ion "gates" set along the membrane-bound ion channels that control the flow of Na$^+$ and K$^+$. A depolarized cell membrane results in a greater influx of Na$^+$ into the cell. This facilitates electrical impulses across the length of the muscle fiber and continues through the muscle fiber's internal system of membranes and pathways via T-tubules and the sarcoplasmic reticulum (Fig. 4.9).

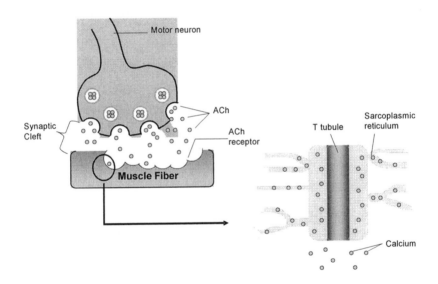

Fig. 4.9. Key players in muscle action.

When an electrical signal arrives at the sarcoplasmic reticulum, calcium (Ca^{2+}) will be released from its usual storage place in the lateral cisternae through calcium release channels called ryanodine receptors. Ca^{2+} is released into the sarcoplasm and immediately binds with troponin-C on the actin filament. Tropinin then moves tropomyosin from covering "active" sites on the actin so that the myosin head can connect with the actin filament. When the myosin and actin connect, tension can be generated. The pulling of actin towards the M line causes concentric (muscle shortening) and the pushing of actin away from the M line causes eccentric (muscle lengthening) contractions. An isometric contraction occurs when the actin and myosin filaments generate force in forming a cross bridge, but the actin filaments do not move toward or away from the M line. Forceful muscle contractions will continue until calcium is pumped back into the sarcoplasmic reticulum where it remains until a new nerve impulse acts on the ryanodine receptors and releases the Ca^{2+} back into the sarcoplasm.

There are many steps in the motor-nerve interactions where radicals are produced. And there are many redox sensitive targets in the muscle cell (Fig. 4.10). Intracellular Ca^{2+} can both stimulate ROS and be stimulated by ROS (Gordeeva *et al.*, 2003). H_2O_2, •O_2-, singlet oxygen, and Ca^{2+} are known to activate calcium channels. Ryanodine receptors contain regulatory thiols (SH) that when oxidized, cause channel opening and Ca^{2+} release. When reduced by sulfhydryl reducing agents such as glutathione (GSH) regulatory thiols on the ryanodine receptors cause channel closure and Ca^{2+} reuptake by ATP-driven Ca^{2+} pumps (Stoyanovsky *et al.*, 1997). For optimal muscle functioning, a favorable redox state must exist on the ryanodine receptor (critical thiol groups should be oxidized) so that Ca^{2+} can be released through the Ca^{2+} channels into the sarcoplasm to initiate actin-myosin cross bridge formation. Then these critical thiol groups should be subsequently reduced so that the channels can be closed (Zaidi *et al.*, 1989). However, a biphasic response has been reported whereby oxidation causes an initial increase in Ca^{2+} gate channels, but further oxidation closes the gate channels (Aghdasi *et al.*, 1997).

Ca^{2+} is not the only component of the contractile muscle apparatus that is closely tied to oxidation-reduction reactions. Every action potential along the sarcolemma is accompanied by •O_2- generation (Bassot *et al.*, 1995) so the potential for oxidative stress during muscle action is substantial. This high level of •O_2- generation may explain why there are over 40 different types of thiols that are contained in actin and myosin filaments. While the thiol groups on actin are targets for oxidation, the high concentrations of GSH and high concentration of GSSG required for oxidizing actin contribute to actin's strong resistance to oxidation (Dalle-Donne *et al.*, 2003). Myosin contains many thiol residues, however, the combination of myoglobin and H_2O_2 exposure has been found to cross link myosin filaments, leaving them vulnerable to oxidative stress, lower Ca^{2+} -ATPase activity, and a significant loss in muscular action (Kamin-Belsky, 1998).

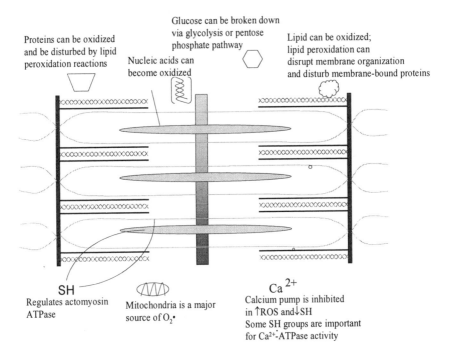

Fig. 4.10. Redox sensitive targets in a muscle cell.

A substantial portion of ATP is catabolized for energy required in muscle contractions. In fact, ATP depletion in cells can be approximated from increased concentrations of extracellular hypoxanthine, a central intermediate in the metabolism of ATP (Harkness and Saugstad, 1997). Hypoxanthine can produce $\cdot O_2$- via the xanthine oxidase pathway. Proteins, lipids, and carbohydrates embedded in the muscle also are affected by and can in turn affect oxidation-reduction reactions.

Oxidative stress can be attenuated by reduced oxygen flux, reduced catecholamine secretion, stable ATP levels (e.g. less energy depletion), and enhanced mitochondrial enzyme activities. All of these are known to occur in exercise trained individuals in rested conditions and when working at sub-maximal intensity levels. Another way to diminish oxidative stress is by upregulating antioxidant defenses. Exercise-induced oxidative stress will induce the activity of some but not all antioxidants.

Even with parallel, serial, and overlapping functions, antioxidants are not 100% effective in preventing the oxidation of lipids, proteins, carbohydrates, and nucleic acids. Some will become end-products that may disrupt cell functioning by their physical presence. Other oxidative stress side and by-products act as cell signalers that initiate reactions that impact gene regulation for repair, growth, development, and turnover. ROS can change the redox status inside a cell and thereby influence a variety of cell signaling pathways that regulate function and viability.

4.8 Aging and Muscle Action

Reduction in muscle strength is a normal consequence of age (Evans, 1995). Both cross sectional and longitudinal data show that muscle strength declines by about 15% per decade in people aged 60 and 70 years (Murray *et al.,* 1985; Harries and Bassey, 1990), mainly due to an age-related decrease in muscle mass. Most studies have reported a selective decline in Type II muscle fibers (approximately 50%) and an increased percentage of Type I fibers with age (Tarpenning *et al.,* 2004; Hunter *et al.,* 1999; Leeuwenburgh, 2003; Oh-Ishi *et al.,* 1995). When a critical threshold of age-related muscle loss is reached, sarcopenia occurs (Fig. 4.11). Sarcopenia is severe muscle atrophy that is a result of a multitude of cellular changes (Fulle *et al.,* 2004). Criteria for sarcopenia are somewhat arbitrary, but an obvious compromise in functional ability occurs with its onset, thus limiting some types of activities of daily living in humans and survival in animals that rely on fast muscle actions for food, shelter, and safety. Oxidative stress is believed to play a major role in sarcopenia (Leeuwenburgh *et al.,* 2003).

Increased ROS can contribute to sacropenia, cell damage and death via two distinct processes: apoptosis and necrosis. Apoptosis is a type of programmed cell death where cells follow an orderly series of self-destructive events that include: cell shrinkage, cell bubbling, and the breakdown of DNA, protein, mitochondria, and cell membrane. The damaged cell membrane attracts phagoyctic cells that engulf the damaged cell membrane fragments and secrete anti-inflammatory cytokines such as IL-10 and TGF-*B*. This is a very different process than necrosis. Necrosis is the death of cells in a tissue usually caused by an injury. Necrosis can be caused by an exercise-induced mechanical or metabolic attack on the cell. As a result of the attack, osmotic pressure and swelling increases inside the cell to a bursting point after which the cell contents actually spill out through ruptures in the membrane. The outpouring of the cell's contents and pro-inflammatory cytokines (e.g. tumor necrosis factor (TNF)-alpha, (interleukin) IL-6) secreted by the cell cause damage to the cell itself and to nearby cells (Fig. 4.12).

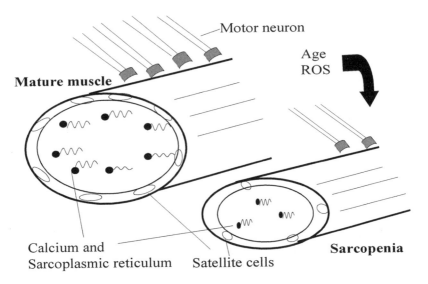

Fig. 4.11. Sarcopenia is represented by fewer and smaller motor neurons, smaller satellite cell pools, less calcium and less extensive sarcoplasmic reticulum system, and smaller sized muscle.

Apoptosis can theoretically benefit an organism by programming damaged or malfunctioning cells to self-destruct in an orderly fashion. But sometimes the program for orderly cell death becomes modified due to injury or a genetic modification (e.g. inactivation of a tumor suppressor gene). This is a highly disordered process. While apoptosis is the preferred method of cell death, both processes require the orchestration of other cells and systems to mop up the fallout from the dead cells. An increased rate of either apoptosis or necrosis results in sarcopenia.

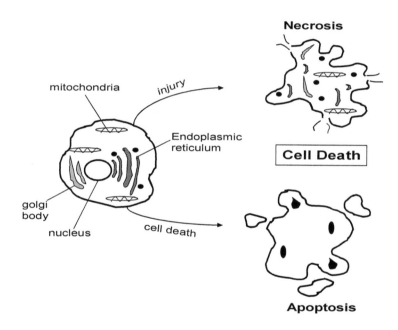

Fig. 4.12. Different types of cell death: Apoptosis and necrosis.

Weindruch (1995) summarized five different mechanisms that may explain a relation between oxidative stress and sarcopenia: (1) age-related increase in lipofuscin, usually caused by the breakdown of damaged blood cells, in muscle; (2) age-related increase in several indices of lipid peroxidation including TBARS and MDA. TBARS are low molecular weight end-products of oxidized lipids. MDA is one of several possible by-products of lipid peroxidation that fall into the TBARS category; (3) age-related increases in the antioxidants GSH, SOD, and CAT (Ji *et al.*, 1990; Leeuwenburgh *et al.*, 1994), probably indicating a compulsory response to free-radical leakage from mitochondria that also tends to rise with age (Lane, 2003); (4) age-related degradation in performance of the mitochondrial electron transport chain (Feuers *et al.*, 1993); and (5) age-related increase in the number of DNA deletions in mitochondria (Chung *et al.*, 1994).

4.9 Summary

Cells have adapted to different forms of oxidative stress over the past 3.5 billion years. Major causes of oxidative stress today include exercise and aging. During exercise, the internal milieu of skeletal muscle is of particular interest due to a combination of internal and external stressors that bring about adaptations immediately and in the hours and days following exercise. Exercise requires an increase in energy use and oxygen uptake. Actions of the neuromuscular system trigger the release of calcium, H^+, heat, and catecholamines (e.g. norepinephrine, epinephrine). While contributing to muscle contractions during exercise, these same conditions also contribute to ROS in the motor neuron and muscle. ROS metabolism in muscle exemplifies pleiotropy because while required for muscle functioning during exercise, these same physiological factors are associated with consequences such as fatigue, overheating, muscle cramping, and oxidative stress. Muscle adaptations that occur in the hours and days following exercise probably include removal and/or repair of exercise-induced damage. That exercise-induced oxidative stress can have beneficial outcomes in most healthy individuals supports the phenomenon of hormesis where metabolic disruption from low to moderate stimulation results in positive biological outcomes. Aging is associated with increased ROS produced in muscle that may be met partly but not completely by increased antioxidant activity. It appears that enhancing defense systems to protect against oxidative stress, by regular exercise or possibly nutrition and supplements are important interventions to prevent or delay sarcopenia.

References

1. Aghdasi, B. *et al., J. Biol. Chem.* **272** (1997), 25462-25467.
2. Alessio, H.M. *et al., Am. J. Physiol.* **255** (1988), C874-7.
3. Alessio, H.M. and Goldfarb, A.H. *J. Appl. Physiol.* **64** (1988), 1333-6.
4. Bassot, M. and Nicolas, M.T. *Histochem. Cell Biol.* **104** (1995), 199-210.
5. Caillaud, C. *et al., Free Radic. Biol. Med.* **26** (1999), 1292-9.
6. Calabrese, E.J. and Baldwin, L.A. *Hum. Exp. Toxicol.* **21** (2002), 91-97.
7. Clarkson, P.M., Tremblay, I. *J. Appl. Physiol.* **65** (1988), 1-6.
8. Costill, D.L. *et al., Intl. J. Sports Med.* **8** (1987), 103-106.
9. Cutler, R.G. *Aging and Cell Function*, ed. Johnson, J.E. Jr. (Plenum Press, New York, 1984), 1-148.
10. Cutler, R.G. *Ann. NY. Acad. Sci.* **621** (1991), 1-28.
11. Dalle-Donne, I. *et al., Free Radic. Biol. Med.* **35** (2003), 1185-93.
12. Evans, W. J. *Clinics Geriat. Med.* **11** (1995), 725-34.
13. Fridovich, I. and Freeman, B. *Ann. Rev. Physiol.* **48** (1986), 693-702.
14. Fulle, S. *et al., Exper. Geronolt.* **39** (2004 Jan), 17-24.
15. Gordeeva, A.V. *et al., Biochem. Biokhimiia* **68** (2003), 1077-80.
16. Harkness, R.A. and Saugstad, O.D. *Scand. J. Clin. Lab. Invest.* **57** (1997), 655-72.
17. Harries, U.J. and Bassey, E.J. *Eur. J. Appl. Physiol. Occup. Physiol.* **60** (1990), 187-90.
18. Huang, J. and Forsberg, N.E. *Proc. Natl. Acad. Sci.* **95** (1998), 12100-5.
19. Hunter S.K. *et al., J. Appl. Physiol.* **86** (1999), 1858-65.
20. Jagoe, R.T. and Goldberg, A. L. *Curr. Opin. Clin. Nutr. Metab. Care* **4** (2001), 183-90.
21. Ji, L.L, *et al., J. Appl Physiol.* **73** (1992), 1854-1859.
22. Ji, L.L. *Med. Sci. Sports Exerc.* **25** (1993), 225-231.
23. Kamin-Belsky, N. *et al., Adv. Exp. Med. Biol.* **454** (1998), 219-23.
24. Kirkendall, D. T. and Garrett, W.E. Jr. *Am. J. Sports Med.* **26**(1998), 598-602.
25. Lawler, J. M. *et al., Am. J. Physiol.* **265** (1993), R1344-50.
26. Leeuwenburgh, C. *J. Gerontol. A Biol. Sci. Med. Sci.* **58** (2003), 999-1001.
27. Moore, J. D. *et al., J. Pharmacol.***135** (2002), 1069-77.
28. Murray, M.P. *et al., J. Gerontol.* **40** (1985), 275-80.
29. Oh-Ishi, S. *et al., Mech. Age. Dev.* **40** (1985), 275-280.
30. Reid, M.B. and Durham, W.J. *Ann. N Y Acad. Sci.* **959** (2002 Apr), 108-16.
31. Roy, R. *et al., J. Appl. Physiol.* **59** (1985): 639-46.

32. Saul, R.L. *et al., Modern Biological Theories of Aging*, ed. Warner, H.R. (Raven Press, New York, 1987), 113-129.
33. Sen, C.K. and Packer, L. *Am. J. Clin. Nutr.* **72**, (2000 Aug), 653S-69S.
34. Orr, W.C. and Sohal, R. S. *Science.* **263** (1994), 1128-30.
35. Staron, R.S. *Can. J. Appl.Physiol.* **22** (1997), 307-327.
36. Stevenson, E.J. *et al., J. Physiol.* **551** (2003), 33-48.
37. Stoyanovsky, D. *et al., Cell calcium.* **21** (1997), 19-29.
38. Tarpenning, K.M. *et al., Med. Sci. Sports Exerc.* **36** (2004), 74-8.
39. Weindruch, R. *J. Gerontol. A Biol. Sci. Med. Sci.* **50** (1995), 157-61.
40. Yu, B.P. and Yang, R. *Ann. N Y Acad. Sci.* **786** (1996), 1-11.
41. Zaidi, N.F. *et al., J. Biol. Chem.* **264** (1989), 21725-36.

CHAPTER 5

OXIDATIVE STRESS ACROSS THE EXERCISE CONTINUUM

H.M. Alessio

Miami University, Oxford, OH

5.1 Introduction

During most exercises and physical activities, a combination of isometric, concentric, and eccentric muscle contractions generate force. In doing so, all muscle contractions across the exercise continuum produce some amount of mechanical and/or metabolic stress. In his review of over one hundred studies on exercise-induced oxidative stress Pyne (1994) described how mechanical and metabolic stress cause muscle damage. Evidence of muscle damage following novel types of exercise, eccentric muscle contractions, and high-intensity exercise using isometric, eccentric or concentric contractions have been reported in many studies. Electron microscopy studies have reported Z-line breaks (Belcastro *et al.,* 1998), biomechanical studies have reported reduced range of motion in the affected joints (Dutto *et al.,* 2004), biochemical studies have reported increased creatine kinase (CK) (Clarkson, 1997), 3-methyl-hystidine (Chevion *et al.,* 2003), pro-inflammatory cytokines (Chan *et al.,* 2004), tumor necrosis factor (TNF) (Horne *et al.,* 1997), interleukins (Ostrowski *et al.,* 1998), and many studies have reported muscle soreness and fatigue using a variety of scales (Thompson *et al.,* 1999; Clarkson and Newham, 1995).

5.2 Muscle Contractions and Reactive Oxygen Species

Mechanical and metabolic stress associated with virtually any type of exercise will contribute to prooxidant activity by increasing reactive oxygen species (ROS) by one or more of a number of different pathways. That is because all types of muscle contractions across the exercise continuum require action potentials of varying amounts and magnitude. Every action potential along the muscle cell membrane is accompanied by $\cdot O_2^-$ generation (Bassot *et al.*, 1995). High intensity muscle contractions in particular, require continuous depolarization across the sarcolemma, followed by movement of calcium out of the sarcoplasmic reticulum to facilitate cross bridge formation and tension. Excessive intracellular calcium activates calcium dependent proteases that convert xanthine dehydrogenase to xanthine oxidase. These reactions produce $\cdot O_2^-$ (Jackson and O'Farrell, 1993).

Both concentric and eccentric exercise experience increased mitochondrial respiration, which increases vulnerability for ROS production. Eccentric muscle contractions however, are more likely than concentric to cause physical damage to the sarcomere, specifically in myofibrillar lesions along the Z-lines and alterations in the architectural proteins desmin, nebulin, and α-actinin (Yu *et al.*, 2003). Muscle damage induced by eccentric exercise is often accompanied by evidence of oxidative stress, muscle soreness, and reduced muscle function. Nevertheless the alterations in these architectural proteins are key steps in muscle remodeling that takes place as part of muscle adaptation. Lee *et al.* (2002) examined the effects one bout of high-intensity eccentric exercise on oxidative stress markers including blood protein carbonyls and GSH status and muscle damage indicators CK and muscle soreness. They reported a 900% increase in CK and a 23% reduction in GSH, which were accompanied by a reduction in muscle function and subjective reports of increased muscle soreness.

5.3 Age and Exercise-Induced ROS

Age could be a factor in eccentric-exercise induced oxidative stress because of the potential for more ROS produced from a mitochondrial respiratory chain that is known to functionally decline with age (Trounce *et al.*, 1989). On the other hand, antioxidant activity tends to increase with age (Leeuwenburgh *et al.*, 1994). An age-related increase in antioxidant activity may not meet the age-related oxidant activity produced by defective, leaky mitochondria. An additional consideration is the compromised immune system of older adults. Phagocytic response to muscle damage caused by eccentric exercise may be inadequate in older compared with younger individuals, and result in oxidative damage to DNA and the accumulation of harmful by-products in around the muscle. Sacheck *et al.* (2003) compared biomarkers of oxidative stress and muscle damage in old and young individuals following eccentric exercise. They found that fitness level was more important than age, in determining protection against exercise-induced oxidative stress and muscle damage. Lipid peroxidation (lipox) as indicated by a prostaglandin-like compound, $F_{2\alpha}$-isoprostane, was directly related to CK. They also reported that vitamin E supplementation did not affect any biomarkers of oxidative stress or damage following eccentric exercise, which indicates that physical damage from exercise cannot be prevented by the biochemical properties of antioxidants.

5.4 ROS in Aerobic and Anaerobic Environments

Exercise-induced oxidative stress in skeletal muscle occurs in both anaerobic and aerobic environments. Anaerobic exercise is usually short term and high intensity. It includes activities such as short duration sprints and power movements that could be part of a more aerobic movement such as throwing or kicking an object after running at submaximal speed. The major portion of anaerobic exercise usually requires a large supply of ATP that is broken down to produce energy. When large amounts of ATP are rapidly degraded, ADP can continue in catabolic reactions to form xanthine and uric acid. Xanthine oxidase catalyzes these reactions and produces $\cdot O_2-$. Anaerobic exercise is often accompanied by increased secretion of the catecholamines norepinephrine and epinephrine. These act as neurotransmitters and first messengers in energy transfer and skeletal muscle actions. Catecholamines can undergo autooxidation with and without oxygen and produce $\bullet O_2-$ (Miller *et al.*, 1996). Lactic acid accumulates during high intensity exercise where oxygen supply often but not always does not meet the demand. Lactic acid can change $\bullet O_2-$ into a more dangerous and highly reactive radical, $OH\bullet$ (deGroot *et al.*, 1995). In an aerobic environment $\bullet O_2-$ can be produced via a mass action effect from oxygen that does not undergo complete reduction to water in the mitochondria. The end result is a small, but significant fraction of $\bullet O_2-$ escapes from the electron transport chain and initiates or propagates harmful radical reactions.

5.5 Muscle Damage and ROS

Exercise can also increase ROS by the physical stress that happens during a wide range of muscle actions across the exercise continuum. Damage to muscle can be caused by physical pressure, muscle and connective tissue tears, removal of growth factors, secretion of glucocorticoids, intracellular Ca^{2+} accumulation, and increase in TNF, which can lead to cell death. When a cell is damaged or destroyed, the body responds by activating the immune system in a similar way that it responds to infection or bacteria. Phagocytic cells, called neutrophils are first-responders to injury or infection. They leave nearby blood vessels and are attracted by the chemical characteristics of the cell damage and newly formed and unrecognizable cell fragments. Upon arriving at the damaged site, the neutrophils act in one of the following ways. They:

1. Produce $\bullet O_2-$ and H_2O_2 to destroy cell damage and fragments
2. Release NADPH oxidase which can produce $\bullet O_2-$ and other ROS
3. Release proteolytic enzymes such as protease and amylase to facilitate a cycle of cell damage and repair

4. Digest cell damage and fragments from the extracellular fluid, deliver them to the lysosome and destroy them with lysosomal enzymes

Nieman has described a "J-shaped model" whereby exercise can enhance or reduce immune function depending in large part on intensity of the exercise (Nieman, 2003). Exercise frequency and duration also play a role in immune function. This model reflects the phenomenon of hormesis where a small dose of something that is potentially harmful or even toxic can result in compensatory responses that ultimately benefit an organism (Calabrese and Baldwin, 2002). This explains how low intensity exercise may cause low level muscle damage that can activate appropriate damage and repair cycles and result in cell turnover and ultimately cell growth and development. On the other hand, high intensity exercise can cause muscle damage and cellular events that activate many different pathways with an end result of cell destruction.

Phaneuf and Leeuwenburgh (2001) described the following exercise-induced factors that contribute to apoptotic cell death:

1. Increased oxidant production in mitochondria
2. Depletion of GSH stores
3. Release of caspase-activating proteins such as cyctochrome c and apoptosis-inducing factor from the mitochondria

5.6 Isometric Exercise and Oxidative Stress

According to the exercise continuum, skeletal muscle contractions fall along a range with isometric at one end and isotonic at the other. In reality, most muscle contractions are somewhere in between the two extremes. Generation of force depends on proper alignment or overlap of actin and myosin filaments. In isometric contractions, no sliding movement occurs between actin and myosin, nevertheless, an active tension exists between the two myofilaments that allows for cross bridges to form and for tension to develop. Despite little to no muscle movement, evidence of oxidative stress has been observed during isometric exercise (Alessio *et al.*, 2000). Causes of oxidative stress

during isometric exercise include elevated intracellular calcium that gets
trapped inside the sarcomere during an isometric contraction,
autooxidation of catecholamines released during intensive isometric
contractions, rapid depletion of ATP stores, lactic acid accumulation,
heat production, redox status of the muscle, and reperfusion following
ischemia during isometric muscle contraction and release.

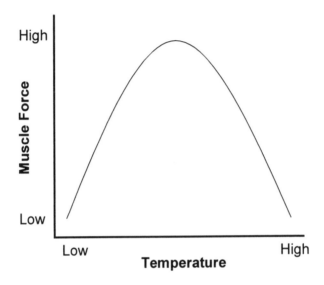

Fig. 5.1. Curvilinear relation between force production and muscle temperature.

Activation of muscle fibers always results in increased heat
production. Unlike isotonic contractions where 75 to 80% of chemical
energy is converted to heat, all excess energy output in isometric
exercise is converted to heat (Hill, 1958). Heat production enhances
contractile activity, which may partly explain why isometric muscle
contractions can generate so much force. Heat production and muscle
force generation are directly related to a point after which the relation
breaks down. Saugen and Vollestad (1995) described a curvilinear
relation between increased temperature and force production during
isometric muscle contractions (Fig. 5.1). The rate of temperature rise
increased with intensity of maximal voluntary contraction (MVC) from
3.1 ± 1.1 mK/s at a low intensity (10% MVC) to 14.5 ± 1.3 mK/s at 90%

MVC. It was in the range of 30 to 70% of MVC that the relation between temperature rise and muscle force was linear. Following 70% MVC, force production leveled off, and further increases in temperature did not enhance force production.

In addition to heat, the redox state of muscle influences maximum isometric force production. Plant *et al.* (2003) explained that rested muscles exist in a slightly reduced state. Force production is low when the muscle is in a reduced state. With increasing exposure to an oxidant, H_2O_2 for example, a redox balance shift occurs and isometric force production increases. The relation between muscle redox state and force production is not linear, however, and when excess oxidants accumulate, contractile proteins can become oxidized and functionally altered. These results were reported by Reid *et al.* (1992) in muscle fiber bundles from rat diaphragm. When subjected to low-frequency contractions that lasted five min, muscle contractions demonstrated an increase in intracellular oxidant levels determined by dichlorofluorsecin, a fluorochrome that emits at a specific wavelength when oxidized. Under conditions of a more oxidized state, muscles became fatigued. Reid *et al.* (1992) also reported that muscle force was improved by addition of the antioxidants catalase (CAT) or superoxide dismutase (SOD).

So, both heat production and redox state in the muscle are directly related to increased isometric force production, however, in both cases, the relationships are not linear. Isometric muscle force will decrease under conditions of excessive heat production. It will also decrease when the muscle contains either very high antioxidant levels or very high oxidant levels.

5.7 Isometric Exercise and ROS

During isometric hand grip exercise, the driving pressure for blood flow into the forearm is lowered even when working at just 10% of one's maximum voluntary contraction (MVC). By 50% MVC the driving pressure of blood flow approaches zero and blood flow into and out of the muscle virtually stops (Sjogaard *et al.*, 1988). When the muscle contraction is released, a reactive hyperemia follows and blood flow through the muscle is restored. Ischemia-reperfusion is another way in which oxidative stress occurs as an occluded tissue with low oxygen levels suddenly becomes saturated with oxygenated blood. The condition of ischemia reperfusion includes two conditions that have been shown to produce $\cdot O_2-$: (1) When the oxygen concentration is greatly increased (during unimpeded blood flow) or (2) when the respiratory chain becomes fully reduced (as happens during ischemia). The ischaemia-reperfusion injury hypothesis was proposed over twenty-five years ago by McCord (1985). He described the way in which energy depletion during ischaemia-which occurs during isometric muscle contraction-eventually changing ATP to hypoxanthine. In between or in recovery from isometric contractions, blood flow is restored by reperfusion. Oxygen delivery to muscles resumes and in the presence of oxygen and xanthine oxidase, hypoxanthine yields xanthine and uric acid. Side products of the two reactions: (1) hypoxanthine to xanthine and (2) xanthine to uric acid, are $\cdot O_2-$ and $\cdot OH$, respectively. This cascade is likely to occur during isometric muscle contractions requiring large amounts of ATP.

Alessio *et al.* (2000) compared biomarkers of oxidative stress in individuals who completed high-intensity isometric exercise and high-intensity aerobic exercise. Both types of exercises were performed for a similar length of time. Aerobic exercise was performed when subjects ran on a treadmill for approximately 15 min following a graded exercise test protocol. Oxygen consumption was tracked and blood samples were collected before, immediately following, and 1 hr following the maximum effort on the treadmill. Subjects also performed isometric muscle contractions that represented 50% of their MVC with their dominant hand. Subjects held the 50% MVC for 45 seconds, then rested for 45 seconds, and repeated this pattern until the accumulated time equaled their time on the treadmill. Oxygen consumption was tracked and blood samples were collected before, immediately following, and 1 hr following the last contraction at the end-time point, which was held until failure. Heart rate increased above rest in both isometric and aerobic exercise, but comparatively more during aerobic exercise (125% in aerobic vs. 25% in isometric exercise). Mean arterial pressure increased above rest in both isometric (40%) and aerobic (10%), but comparatively more during isometric exercise. Oxygen consumption increased 14-fold vs 2-fold in aerobic compared with isometric exercise. Evidence of oxidative stress was seen following both aerobic and isometric exercise, with LH increasing 36% following isometric and 24% during aerobic exercise. Protein carbonyls increased 67% following aerobic exercise and only 12% following isometric exercise (Fig. 5.2).

Since oxygen consumption did not change during isometric exercise, a mechanism other than a mass action effect of oxygen consumption must explain the increase in at least one marker of oxidative stress-LH-since this increased significantly above rest without an increased oxygen uptake. However, ischemia-reperfusion is also a possible contributor to oxidative stress where extremely low oxygen tension during ischemic conditions is followed by excessive oxygen influx during reperfusion in between isometric muscle contractions.

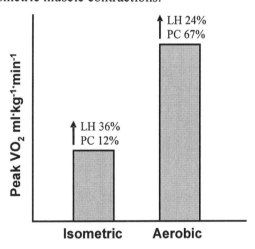

Fig. 5.2. Elevated lipid hydroperoxides (LH) following isometric exercise with low peak VO_2 and elevated protein carbonyls (PC) following aerobic exercise with high peak VO_2.

5.8 Exercise, Hypoxia, and Oxidative Stress

Oxidative stress during isotonic exercise is usually associated with increased VO_2. However, many studies investigating high altitude exposure (Jefferson *et al.*, 2004; Bartsch *et al.*, 2004) and sleep apnea (Xu *et al.*, 2004) have reported evidence of oxidative stress in conditions where oxygen supply is lower than normal. Bartsch *et al.* (2004) reviewed studies investigating pathophysiological systems associated with hypoxia. Results from these studies suggest that oxygen radicals are elevated in hypoxic environments and oxidative stress is involved in the pathophysiology of AMS. Furthermore, antioxidants, particularly GSH can provide some protection (Magalhaes *et al.*, 2004) against hypoxia-induced ROS at rest and following exercise. Bailey *et al.* (2003) separated a group of subjects into either a normoxic (O_2= 21%) or a hypoxic (O_2=16%) group and exercised them all to exhaustion on a bicycle ergometer. Blood levels of LH and malondialdehyde (MDA) increased in subjects who exercised under hypoxic conditions, despite a lower maximal VO_2max compared with normoxic conditions. Increases in LH and MDA were correlated with the exercise-induced decrease in arterial hemoglobin oxygen saturation (r = -0.61 and r = -0.50 respectively), but not with VO_2max. In fact, the authors reported a disassociation between VO_2 and radical production as measured by electron spin resonance (ESR). It appeared that exercise-induced oxidative stress was more intimately regulated by increased mitochondrial redox subsequent to a decrease in mitochondrial PO_2 rather than the traditionally held hypothesis of increased PO_2 and electron flux. This paradoxical finding provided support for factors other than increased VO_2 that could increase oxidative stress, such as H+ generation, norepinephrine autooxidation, peroxidation of damaged tissue, and xanthine oxidase activation (Bailey *et al.*, 2003). Bailey and his colleagues concluded that exercise-induced oxidative stress (1) can increase during hypoxia; (2) is not regulated exclusively by a mass action effect of VO_2; and (3) is selectively attenuated by regular hypoxic training.

5.9 Resistance Exercise and Oxidative Stress

Moving along the exercise continuum from isometric towards isotonic are several types of exercises that involve high-intensity muscle contractions that generate a lot of force over a short period of time. Exercise induced oxidative stress during resistance exercise has not been studied as extensively during aerobic exercise. One of the first studies by McBride *et al.* (1998) compared MDA and CK following high intensity resistance exercise in recreational lifters without and with two weeks of vitamin E antioxidant supplementation. Each subject performed eight different resistance exercises in a circuit fashion. A total of three sets of 10 maximum repetitions for each set was performed to reach exercise with a two min rest between each set on the first circuit, a 1.5 min rest for the second circuit and a one min rest for the remaining circuits. Plasma MDA significantly increased following the high intensity resistance exercise circuit in subjects with and without antioxidant supplementation. In both conditions, MDA reached a similar peak level that was approximately 100% above resting, but at different time points. MDA peaked immediately following resistance exercise in the supplement group and then decreased towards baseline levels six hr later. In the group that did not take supplements, MDA peaked six hr following resistance exercise and remained at a similar level up to 24 hr later. It appeared that vitamin E assisted in clearing out MDA faster than the group that did not take antioxidants.

Avery *et al.* (2003) compared markers of oxidative stress, muscle damage, and muscle soreness in recreational subjects that participated in resistance exercise that included concentric and eccentric muscle contractions in all of the major muscle groups. Half of the subjects took vitamin E supplements and half took a placebo for three weeks. After three weeks, all subjects performed three high intensity resistance exercise sessions separated by three recovery days. Peak strength occurred on day one. Muscle soreness peaked after the first day of resistance exercise and MDA peaked a week later. The authors concluded that acute resistance exercise can increase biomarkers of oxidative stress, muscle soreness, and muscle damage. Vitamin E supplementation did not affect any of these markers.

Vincent *et al.* (2002) investigated whether chronic resistance exercise may act in a similar way as chronic aerobic exercise, and contribute to endogenous factors that enhance antioxidant defense and/or reduce pro-oxidant attack. They studied 62 men and women who participated in six months of resistance training and compared biomarkers of oxidative stress following acute aerobic exercise. They reported that resistance training reduced serum lipox following aerobic exercise and provided protection against oxidizing agents *in vitro*. They concluded that resistance training provided a "cross-protection" against aerobic-exercise induced oxidative stress and they attributed this protection to increased GSH levels.

5.10 Sprint Exercise and Oxidative Stress

Sprinting is an example of a predominantly anaerobic exercise where the majority of energy transfer is derived from the phosphagen system with little dependence on oxygen flux. Sprinting shares some characteristics of isometric exercise although there is clearly muscle movement involved in sprinting. Some characteristics shared by both isometric and sprinting exercises include high power output and generation of force, low dependence on oxygen for ATP, persistent action potential across the sarcolemma, rapid degradation of ATP that may continue along the xanthine oxidase pathway, increased heat production, increased intracellular calcium levels, and increased acid levels. Sprinting exercise has been associated with enhanced levels of lipox and other byproducts of ROS (Alessio *et al.*, 1988; Atalay *et al.*, 1996). Atalay *et al.* (1996) also reported that 6 weeks of sprint training resulted in increased muscle GSH, glutathione S-transferase, and glutathione reductase in Type II muscle fibers and heart muscle of rats.

Marzatico *et al.* (1997) measured plasma indices of lipox, conjugated dienes (CD) and MDA and erythrocyte enzymes SOD, GPx and CAT in sprinters, marathon runners, and non-runners. Compared with non-runners, resting MDA was higher in both sprinters and marathon runners. This result suggests that runners-in-training, regardless whether they are sprinters or distance runners, are under more oxidative stress than non-runners. Marathon runners also had higher CD than non-runners. Two antioxixants, SOD and GSHPx were elevated in sprinters indicating an up-regulated enzymatic scavenging capacity, although one antioxidant, CAT was lower in runners compared with non-runners. These antioxidant enzymes did not change following anaerobic exercise. Nevertheless, CD peaked 6 hrs following sprinting and MDA gradually increased until 48 hrs after sprinting. Marzatico's study implies that modifications to sprinters' antioxidant defense occur to address prooxidant stress associated with sprinting, but protection is not certain.

Using an animal model, Kayatekin *et al.* (2002) had mice perform 15 bouts of sprinting exercise, each comprising running on a treadmill for 30 s at 35 m/min up a 5° slope, with a 10 second rest interval between bouts. Muscle TBARS increased at 0.5 and three hr post-exercise, and then returned to control levels by 24 hr post-exercise. They concluded that acute sprint exercise in mice caused TBARS levels to increase in skeletal muscle but antioxidant enzyme activities were not affected. Here again, the antioxidant response was not able to protect muscles from exercise-induced oxidative stress associated with high intensity sprinting.

A type of exercise similar to running a sprint is fast pedaling on a bicycle ergometer. The 30 second Wingate test is a popular measure of lower body power. It requires a maximum effort where an individual pedals as fast as possible against a force on the flywheel that ranges from three to six kg depending on body weight. Ashton *et al.* (1999) measured radical concentrations directly by ESR in addition to several different biochemical markers of oxidative stress following a 30 second Wingate bicycle test. They reported a 100% increase in the ESR signal following the Wingate exercise. An increase in this radical signal was accompanied by markers of oxidized lipids in the form of MDA (+27%). No significant changes in other markers including LH, ascorbic acid, and alpha tocopherol were observed following the 30 second all-out exercise on a bicycle ergometer.

5.11 Contact Sports and Oxidative Stress

Some contact sports such as football and rugby have isometric and short-term high intensity action components. Short-term high intensity running and hitting usually results in evidence of oxidative stress. The significance of oxidative stress in contact sports may be its association with fatigue and muscle damage. Fatigue and muscle damage can be considered as defense mechanisms that protect muscle from overuse or consequences of injury. A study on oxidative stress in American professional football players showed that several markers of oxidative stress increased from pre-season levels to three different in-season time points. Serum total peroxide concentrations, auto-antibody titers against oxidized low-density lipoprotein (oLab), lag time of ROS-induced degradation of the fluorophore DPHPC and serum ascorbate, alpha- and gamma-tocopherol, and beta-carotene concentrations were measured in 8 professional football players. Before the competitive season, serum antioxidant concentrations were within the lower normal range. During the competitive season, plasma ascorbate concentrations increased possibly in response to increased serum peroxide concentrations, which also increased during the competitive season. The oLab titers increased significantly at the mid-competition period time-point but did not increase further. While none of the athletes showed signs of oxidative stress prior to the football season, half demonstrated evidence for significant increases in oxidative stress during the competitive season. These results suggest that high intensity exercise associated with American football results in oxidative stress. While the specific mechanisms are not identified, they probably include one or more of the following: exercise-induced injury, activation of neutrophils and NADPH oxidase, excessive calcium activation, and ischemia reperfusion during forceful muscle contractions. It is not known if the oxidative stress demonstrated by the football players represented muscle fatigue or muscle damage. It is also not known if the other half of the football players who did not show evidence of elevated oxidative stress had better antioxidant protection, were more fit, or suffered less muscle damage related to the contact sport.

Chang *et al.* (2002) approached some of these questions raised from the American football study in a study of rugby players. They compared well-trained rugby players with recreational rugby players. The trained rugby players still showed signs of oxidative stress before and after a rugby match, however the recreational rugby players had higher very low density lipoproteins + low density lipoproteins-CD and TBARS levels that were almost twice as high as well-trained following a match. Both football and rugby athletes experience oxidative stress as a result of their sports. While the high fitness level of professional and well-trained athletes helps to control oxidative stress, there still is variability in oxidative stress biomarkers among athletes. Nevertheless, the intensity of force produced by muscles, trauma, and metabolic demands of football and rugby can overwhelm endogenous defenses.

5.12 Aerobic Exercise and Oxidative Stress

Isotonic exercise includes rhythmic muscle contractions that rely predominantly on oxygen for energy transfer. The body's major physiological systems work to meet the demands of aerobic exercise by delivering oxygen via increased respiration, heart rate, blood flow, reduction of oxygen in the inner membrane of the mitochondria and production of large amounts of energy via oxidative phosphorylation. The ability of the lungs, heart, blood, and muscle mitochondria work towards "handling" elevated oxygen levels during aerobic exercise.

At the farthest end of the exercise continuum, representing the extreme of aerobic/rhythmic exercise where oxygen handling is critical to performance is ultra marathon and marathon running. Many studies have investigated pro-oxidant and antioxidant biomarkers in marathon and ultra marathon runners. During these races, athletes can perform at up to 85% of their VO_2 max for several hours during a marathon, and several hours over several days during an ultra-marathon. A mass action effect of oxygen probably plays a major role in oxidative stress reactions during marathon and ultra marathon running.

Evidence for oxidative stress during and after ultra marathon races include a 33% increase in $F_{2\alpha}$-isoprostane and 88% increase in LH (Nieman *et al.*, 2003). Radak *et al.* (2003) reported increased serum nitrotyrosine and protein carbonyl levels after the first (93 km) day running which then leveled off on the second (120 km), third (56 km) and forth (59 km) days of a super marathon competition. They also reported that a large percentage of urinary proteins were carbonylated and nitrated. While these increases were statistically different from resting levels and were also probably biologically relevant, it is quite possible the training status of these ultra marathon runners resulted in enhanced antioxidant protection. Elosua *et al.* (2003) reported that 16 weeks of regular exercise increased endogenous antioxidant activity with a 28% increase in GPX and an 18% increase in plasma glutathione reductase. These antioxidant increases were accompanied by LDL that was more resilient to oxidation. Miyazaki *et al.* (2001) described how high intensity endurance training resulted in a reduction in exercise-induced lipox and the antioxidant enzymes GPX and SOD at rest. Exhaustive exercise, however, did not result in a change in antioxidant activity. There are several reasons why studies such as this one have not reported increased antioxidants following long distance exercise: (1) Some antioxidants are actually depleted after "doing their job" to either reduce or remove pro-oxidants or radicals; and (2) A down regulation (Ji *et al.*, 1988) of antioxidants has been shown to occur following training which represented less need for antioxidants due to enhanced muscle oxidative and oxygen handling capacities.

Child *et al.* (1998) compared indices of antioxidant status, membrane permeability, and lipox in seventeen trained runners that completed a self-paced half-marathon run on a motorized treadmill. Their pace averaged approximately 77% of VO_2 max for about 87 min. After the race, increases were observed in serum MDA (11%), total antioxidant capacity (18%), uric acid (16%), cortisol (64%), and CK (36%). These represented increased oxidative stress, antioxidant protection, and muscle damage in trained runners that completed a half marathon. One again, the performance of very long distance exercise resulted in oxidative stress that appeared to overwhelm endogenous antioxidant and other protective defenses.

A question remains as to whether oxidative stress is more sensitive to duration or intensity in endurance exercise. A study by Niess *et al.* (1996) compared markers of DNA damage in subjects who completed a graded exercise test to exhaustion. They reported a significant increase of DNA migration in trained individuals that was half the level of untrained individuals. Different measures of DNA damage have been used in exercise studies and they do not always concur. For example, Pilger *et al.* (1997) studied urinary excretion of a marker of DNA damage, 8-OHdG in a group of trained long distance runners and a group of healthy controls. They did not observe any differences in mean excretion levels of 8-OHdG between runners and controls. Asami *et al.* (1998) used an animal model and compared forced and spontaneous running exercise on 8-OHdG. Forced exercise represents a high-intensity running pace that the animal does not choose, whereas spontaneous running represents a lower-intensity that is likely to be performed over a longer distance in part because the animal controls the pace and the pace is less intense. Levels of 8-OHdG in the heart, lung, and liver DNA were 90%, 110%, and 140% higher in animals forced to run one hr at a high-intensity pace compared with spontaneous wheel runners who had free access to a running wheel. Distance run was not correlated to 8-OHdG levels in any of the organs. The authors concluded that intensity of exercise, not duration, is an important determinant in DNA damage. This conclusion does not mean that duration is not a factor in oxidative stress. Many studies have reported various types of oxidative stress with prolonged exercise. Inayama *et al.* (1996) reported that prolonged exercise results in the oxidation of plasma protein and Vasankari *et al.* (1995) reported elevated CD, which may the most sensitive method to estimate serum lipox with prolonged exercise. Nevertheless, most studies agree that intensity is a critical factor in exercise-induced oxidative stress (McBride and Kraemer, 1999; Nieman and Pedersen, 1999).

5.13 Cross-Training Sports and Oxidative Stress

Basketball and soccer are examples of sports that depend on muscle contractions across the exercise continuum. In basketball, the power involved in vertical jumps for rebounds and endurance necessary for intermittent running over 45 min can cause oxidative stress via a variety of pathways. Soccer players can run up to 10 miles in a 90 min soccer match. Sprinting, kicking, and lunging require short bursts of power and strength. A study by Szczesniak *et al.* (1998) compared biomarkers of oxidative stress in professional basketball players following a treadmill exercise test and reported increased TBARS and decreased GSH. These indices of oxidative stress suggest that professional basketball players who run to exhaustion experience oxidative stress despite their high level of cross-training. Schroder *et al.* (2001) tested professional basketball players before and after antioxidant supplementation and found that antioxidants helped to increase endogenous defense and attenuate lipox as evidenced by LH 24 hr following intense basketball training.

Cazzola *et al.* (2003) measured the following biomarkers in 20 professional soccer players: plasma lipoperoxides, peroxidation, uric acid, vitamin C, vitamin E, bilirubin, SOD and GPx. Erythrocyte membrane rigidity was also measured as evidence of cell damage. Compared to sedentary controls, professional soccer players had improved antioxidant status together with a more fluid membrane status, which could contribute to improving many functional metabolic interchanges in the cellular membrane. Brites *et al.* (1999) found professional soccer players to have higher antioxidant levels and lower levels of low-density lipoprotein oxidation. Another study compared plasma MDA levels and erythrocyte SOD activity in 25 young male soccer players with 25 sedentary controls (Metin *et al.*, 2003). In addition to lower lipox and higher SOD in the soccer players, they also showed a significant correlation ($r=0.42$) between VO_2 max and SOD, which supports a previous finding (Jenkins *et al.*, 1990).

Compared to weight bearing sports on land, swimming is a sport that is usually associated with less physical trauma on muscles and joints and includes continuous and rhythmic eccentric and concentric muscle contractions. Immersion in cool water and controlled breathing changes the kinetics of oxygen uptake, so that VO_2 is generally lower in swimmers working at a near maximal intensity in the water compared with a similar intensity on land.

Radak *et al.*, (1999) compared the effects of long-term swimming training on the oxidative status of phospholipids, proteins, and DNA in 2 age groups of rats, young (4 week) and middle aged (14 months). The main differences were that DNA damage as indicated by 8-OHdG and extent of carbonylation in specific protein of molecular weight around 29 KDa were smaller in the swim-trained compared to the sedentary animals both at rest and following exercise. Moderate swimming appeared to provide protection against some markers of oxidative stress. This result is in contrast to a report by Turgut *et al.* (2003). They found that 30 min of swimming exercise caused increases in liver and heart MDA, which was higher in swim trained versus sedentary animals. Swimming has been shown to result in similar oxidative stress changes as land exercise. Di Simplicio *et al.* (1997) and Alessio *et al.* (2005) reported GSH depletion and increased GSSG following acute swimming to exhaustion. Simplicio also reported a decrease in SOD activity following exhaustive swimming. The extent of oxidative stress due to swimming has varied among studies. Leichtweis *et al.* (1997) reported that high intensity swimming impaired heart mitochondrial function, making them more susceptible to *in vivo* and *in vitro* oxidative stress. They attributed this increased oxidative stress exposure to a diminished GSH reserve. However, GSH depletion may be compensated by other antioxidants during swimming exercise. Terblanche (2000) reported that swimming training in rats increased the activity levels of CAT in the various tissues investigated for both male and female rats an average of 417% (males 404%; females 430%). This suggests that the higher activity levels of CAT as a result of regular swimming exercise might be indicative of a compensatory measure to counteract the possible detrimental effects associated with oxidative stress as evidenced by GSH depletion.

Attempts to generalize oxidative stress associated with swimming have been stalled by contrasting results and complicated pathways. Nakao *et al.* (2000) monitored mice in a 6 week swimming training program and reported that except for kidney, which consistently demonstrated increased SOD in all three isozymes measured (extracellular, Cu-Zn, and Mn), the responses in mouse tissues of three SOD isoenzymes to swimming training were all different. They concluded that the kidney was the most sensitive of the organs they measured to adapt to oxidative stress during swimming training, although they were unable to pinpoint the mechanism. To complicate matters further, Varraso *et al.* (2002) reported that young adult male swimmers had higher irritant symptoms and increases in the activities of erythrocyte Cu-Zn SOD and of plasma GPx. They hypothesized that the production of ROS was not only related to the number of hours of swimming training but also to exposure to chlorinated compounds in the water. They also speculated that while other athletes may have respiratory problems such as asthma, the exposure to chlorinated compounds may put swimmers at increased risk of respiratory illness.

Most exercises include a combination of concentric and eccentric actions that contribute to metabolic and mechanical stress. Many of the classic studies have used animal models in order to measure biomarkers of muscle damage and oxidative stress in blood, muscle, liver, kidney, lungs, heart, and brain. Sen (1995), Alessio (1993) and others have summarized results from decades of studies on oxidative stress. Knowledge about basic mechanisms that produce and protect against exercise induced oxidative stress has provided guidance on intensity, pacing, training, nutrition, body composition, and vitamin supplementation for highly trained and recreational athletes.

Table 5.1 includes a list and brief description of research studies that have investigated oxidative stress in specific sports. Most sports include muscle movements that cover the majority muscle actions of the exercise continuum. Some, like weight lifting and resistance training lean more toward the isometric end of the continuum, while others such as running and soccer lean more towards the rhythmic or isotonic end of the continuum. In general, most studies of different sports agree that trained athletes have superior antioxidant defense systems in place that can prevent the accumulation of oxidative stress biomarkers during submaximal exercise. However, during and following maximal exercise efforts, even the most trained athletes experience signs and symptoms of metabolic and mechanical stress in the form of oxidative stress by-product accumulation (e.g. increased MDA, LH, 8-OH-dG, CD, TBARS, PC) or muscle damage (e.g. CK, 3-methyl histidine, muscle soreness).

Table 5.1. Oxidative stress biomarkers measured during or following different sports in humans.

Sport	Major findings concerning oxidative stress	Reference
American Football	Pre-season, serum antioxidants were in low to normal range. During season, only half of the subjects had evidence of higher serum peroxide levels.	Schippinger *et al.*, 2002
Basketball	TBARS increased and GSH decreased following exhaustive treadmill exercise.	Szczesniak *et al.*, 1998
	Antioxidant supplements in professional basketball players increased serum alpha toc and beta carotene, and reduced lipid peroxides.	Schroder *et al.*, 2001
Soccer	Soccer athletes had reduced lipox and improved SOD activity.	Metin *et al.*, 2003
	Soccer athletes had improved ascorbic acid, uric acid, alpha toc, SOD, and HDL.	Brites *et al.*, 1999
Rugby	Trained compared with recreational rugby players had lower TBARS and CD.	Chang *et al.*, 2002
Running	After a three month running training program, uric acid, SH-groups, alpha-tocopherol, beta-carotene, retinol, but not ascorbate decreased significantly.	Bergholm *et al.*, 1999
	Measured nitroxide, via EPR in trained and untrained. While at rest untrained may benefit more from antioxidant supplementation than trained; conversely, during exercise trained athletes may benefit more from supplementation than untrained.	Paolini *et al.*, 2003
	Endurance training (running or swimming) decreased extracellular-SOD level at rest 22.2%. Acute exercise after the training, but not before the training, increased plasma lipid peroxide level.	Ookawara *et al.*, 2003
	Eight days of high intensity running in trained runners resulted in increased urinary 8-OHdG excretion and increased TBARS, LDH, CK, CK-MB, and myoglobin. Plasma beta-carotene and alpha-tocopherol increased after camp.	Okamura *et al.*, 1997
	Regular running exercise in untrained healthy adults resulted in no significant difference in the mean excretion levels of O-OhdG between runners and control.	Pilger *et al.*, 1997
	DNA damage in white blood cells was reported following exhaustive exercise. Plasma MDA did not increase in trained and untrained after exercise. At rest and 15 min after exercise MDA values were significantly lower in TR compared to UT.	Niess *et al.*, 1996

Half-marathon	After a self-paced half-marathon run on a treadmill MDA, TAC, and uric acid increased 12%, 19%, 16%, respectively. Lipid peroxidation and muscle damage (increased CK) was reported after the run.	Child *et al.*, 1998
	Trained marathon runners completed a half-marathon. Plasma CK increased to a maximum 24 hr after the race but this was not accompanied by changes in lipid peroxidation as evidenced by conjugated dienes and TBARS.	Duthie *et al.*, 1990
Marathon	Following a 42 km marathon race oxidative DNA damage correlated significantly with plasma levels of creatinine kinase and lipid peroxidation metabolites, and lasted for more than 1 week following the race.	Tsai *et al.*, 2001
	Following a marathon race, plasma protein-bound sulfhydryl group values (-22%, $p < 0.01$), and 24 hr (-12%, $p < 0.01$) and 48 h (-13%). Plasma TBARS were unchanged following the race, and creatine kinase increased.	Inayama *et al.*, 1996
Ultra marathon	Runners in a 160 km ultra marathon race had increased post-race indicators of oxidative stress: F (2)-isoprostane and lipid hydroperoxides, increased 33% and 88%, respectively.	Nieman *et al.*, 2003
	Serum nitrotyrosine and protein carbonyl levels increased after the first (93 km) day running and leveled off on the second (120 km), third (56 km) and fourth (59 km) days of a super marathon competition. A large percentage of urinary proteins were carbonylated and nitrated.	Radak *et al.*, 2003
	16 weeks of regular exercise increases endogenous antioxidant activity (GSH-Px (27.7%), P-GR (17.6%), and increases LDL resistance to oxidation.	Elosua *et al.*, 2003
	High intensity endurance training resulted in a reduction in exercise-induced lipid peroxidation in erythrocyte membrane and increased resting levels of SOD and GPX activities. However, there was no evidence that exhausting exercise enhanced the levels of any antioxidant enzyme activity.	Miyazaki *et al.*, 2001
Sprinting	Sprinters had higher SOD and GSHPx, but lower CAT compared with controls. Exhaustive sprint training overwhelmed their capacity to detoxify ROS, producing oxidative stress.	Marzatico *et al*, 1997

Swimming	Swimmers who train in cold water increased their baseline concentration of GSH and the activities of erythrocytic SOD and Cat.	Siems *et al.*, 1999.
	Swimmers had increased Cu2+/Zn2+ SOD activity and decreased erythrocyte GSH-Px as a result of both exercise training and exposure to chlorinated compounds.	Varraso *et al.*, 2002
Triathlon	In rest conditions, overload training in triathletes compromised antioxidant defense as evidenced by increased plasma GSH-Px activity and decreased plasma total antioxidants. In exercise conditions, overload training resulted in higher exercise-induced variations of blood GSH/GSSG ratio, TBARS level, and CK-MB mass and decreased total antioxidant response.	Palazzetti *et al.*, 2003
Weight Lifting and Resistance Exercise	Biomarkers of oxidative stress were measured by following three weeks of resistance training with and without antioxidant supplementation. Plasma MDA was significantly elevated on days 7 and 8. Antioxidant supplementation did not change muscle soreness, performance measures, or plasma MDA.	Avery *et al.*, 2003
	Resting levels of alpha-toc, gamma-toc, beta-carotene, lycopene, ascorbic acid, MDA and CD concentrations did not differ between groups at rest. Resistance training lowered CD during exercise.	Ramel *et al.*, 2004
	Plasma MDA increased following resistance exercise. Vitamin E supplementation resulted in a faster return of MDA to baseline levels and less disruptions on muscle cell membranes.	McBride *et al.*, 1998
	Moderate intensity, weight resistance exercise, despite inducing mild inflammation, depressed plasma serum peroxide levels, especially when combined with 4 weeks of soy consumption.	Hill *et al.*, 2004
	Six months of resistance exercise training reduced serum lipox, provided protection against oxidizing agents *in vitro*, and provided a "cross-protection" against the oxidative stress generated by aerobic exercise, as evidenced by improvements in the thiol portion of the antioxidant defense.	Vincent *et al.*, 2002
	Protein carbonyls increased 12% and LH increased 36% pre- to immediately post-isometric exercise. ORAC increased 9% following isometric exercise.	Alessio *et al.*, 2000

H.M. Alessio

5.14 Summary

There are currently thousands of studies that have investigated exercise-induced oxidative stress. Many of these studies reported increased oxidants (e.g., ROS, RNS, $\cdot O_2$-, OH$\cdot$, LOO$\cdot$, LH, CD, MDA, TBARS, PC, 8-OHdG) are associated with fatigue and muscle damage. Nevertheless, low levels of these same oxidants may act as cell signaling molecules that act to repair damaged cells and promote muscle development and hypertrophy which are important for peak sports performance. A majority of studies investigating exercise and oxidative stress conclude that endogenous antioxidant defense systems are not able to fully defend against ROS produced during high intensity exercise associated with isometric exercise, weight lifting, sprinting, bicycling, running, and sports play. This implies that oxidative stress is potentially harmful and should be avoided whenever possible. It seems contradictory but oxidants are also essential for cell turnover and gene expression for proteins that contribute to muscle function. While universal recommendations specifying types and dosages of antioxidants are difficult to make, Sen (2001) suggests that competitive athletes routinely engaged in high intensity exercise determine their own needs for a balanced diet that contains adequate antioxidants and specific antioxidant supplements. High levels of antioxidants, which may be desirable for cell protection, may however leave muscle cells in a highly reduced state that hinders force development and this may be contraindicated to athletic performance. The different radical producing pathways appear to be modified (usually attenuated) by chronic exercise training-whether it be isometric, isotonic or another exercise along the continuum. However, exercise training alone, does not provide adequate protection against exercise-induced oxidative stress if individuals are untrained or if the exercise is of very high intensity or long duration.

References

1. Alessio, H.M. *et al.*, *Med. Sci. Sports Exerc.* **25** (1993), 218-224.
2. Alessio, H.M. *et al.*, *Med. Sci. Sports Exerc.* **32** (2000), 1576-81.

3. Alessio, H.M. *et al., Physiol. Behav.* **84** (2005), 65-72.
4. Asami, S. *et al., Biochem. Biophys. Res. Comm.* **243** (1998), 678-82.
5. Ashton, T. *et al., J. Appl. Physiol.* **87** (1999), 2032-6.
6. Avery, N.G. *et al., J. Strength Conditioning Res.* **17** (2003), 801-9.
7. Bailey, D.M. *et al., Clin. Sci. (London England: 1979)* **101** (2001), 465-75.
8. Bailey, D.M. *et al., J. Appl. Physiol.* **94** (2003), 1714-8.
9. Bartsch, P. *et al., High Altitude Medicine and Biology.* **5** (2004), 110-24.
10. Bassot, J.M. and Nicolas, M.T. *Histochem.Cell Boil.* **104** (1995), 199-210.
11. Belcastro, A.N. *et al., Mol. Cell. Biochem.* **179** (1998), 135-45.
12. Bergholm, R. *et al., Atherosclerosis.* **145** (1999), 341-9.
13. Brites, F.D. *et al., Clin. Sci. (London England: 1979).* **96** (1999), 381-5.
14. Cazzola, R. *et al., Eur. J. Clin. Invest.* **33** (2003), 924-30.
15. Chan, M.H. *et al., Amer. J. Physiol.* **287** (2004), R322-7.
16. Chang, C. *et al., Ann. Nutr. Metab.* **46** (2002), 103-7.
17. Chevion, S. *et al., Proc. Natl. Acad. Sci. USA.* **100** (2003): 5119-23.
18. Child, R.B. *et al., Med. Sci. Sports Exerc.* **30** (1998), 1603-7.
19. Clarkson, P.M. and Newham, D.J. *Adv. Exp. Med. Biol.* **384** (1995), 457-69.
20. Clarkson, P.M. *Int. J. Sports Med.* **18** (1997), S314-7.
21. de Groot, M.J. *et al., Mol. Cell. Biochem.* **146** (1995), 147-55.
22. Di Simplicio, P. *et al., Eur. J. Appl. Physiol.* **76** (1997): 302-7.
23. Duthie, G.G. *et al., Arch. Biochem. Biophys.* **282** (1990), 78-83.
24. Dutto, D.J. *et al., Med. Sci. Sports Exerc.* **36** (2004), 560-6.
25. Elosua, R. *et al., Atherosclerosis.* **167** (2003), 327-34.
26. Harkness, R.A. and Saugstad, O.D. *Scand. J. Clin. Lab. Invest.* **57** (1997), 655-72.
27. Hill, A.V. *Proc. R. Soc. (Biol.)* **148** (1958), 397-407.
28. Hill, S. *et al., Int. J. Sport Nutr. Exer. Metabolism.* **14** (2004), 125-32.
29. Horne, L. *et al., Clin. J. Sport Med.* **7** (1997), 247-51.
30. Inayama, T. *et al., Life Sci.* **59** (1996), 573-8.
31. Jackson, M.J. and O'Farrell, S. *Br. Med. Bull.* **49** (1994), 630-641.
32. Jenkins, R.R. *et al., Int. J. Sports Med.* **5** (1984), 11-4.
33. Ji, L.L. *et al., Arch. Biochem. Biophys.* **263** (1988), 137-49.
34. Kayatekin, B.M. *et al., Eur. J Appl. Physiol.* **87** (2002), 141-4.
35. Leeuwenburgh, C. *et al., Amer. J. Physiol.* **267** (1994), R439-R445.
36. Leichtweis, S.B. *et al., Acta physiol. Scand.* **160** (1997), 139-48.
37. Magalhaes, J. *et al., Eur. J. Appl. Physiol.* **91**(2-3) (2004), 185-91.
38. Marzatico, F. *et al., J. Sports Med. Phys. Fitness.* **37** (1997), 235-9.
39. McBride, J.M. and Kraemer, W. *J. Strength Conditioning Res.* **13** (1999), 175-183.
40. McBride, J.M. *et al., Med. Sci. Sports Exerc.* **30** (1998), 67-72.
41. McCord, J.M. *New Engl. J. Med.* **312** (1985), 159-63.
42. Metin, G. *et al., Chin. J. Physiol.* **46** (2003), 35-9.
43. Miller, J.W. *et al., Free Radical Biol. Med.* **21** (1996), 241-9.
44. Miyazaki, H. *et al., Eur. J. Appl. Physiol.* **84** (2001), 1-6.

45. Nakao, C. *et al., J. Appl. Physiol.* **88** (2000), 649-54.
46. Nieman, D.C. and Pedersen, B.K. *Recent Dev. Sports Med. (Auckland NZ).* **27** (1999), 73-80.
47. Nieman, D.C. *et al., Int. J. Sports Med.* **24** (2003), 541-7.
48. Niess, A.M. *et al., Int. J. Sports Med.* **17** (1996), 397-403.
49. Okamura, K. *et al., Free Radical Res.* **26** (1997), 507-14.
50. Ookawara, T. *et al., Free Radical Res.* **37** (2003), 713-9.
51. Ostrowski, K. *et al., J. Physiol.* **508** (1998), 949-53.
52. Palazzetti, S. *et al., Can. J. Appl. Physiol.* **28** (2003), 588-604.
53. Paolini, M. *et al., Free Radical Res.* **37** (2003), 503-8.
54. Pilger, A. *et al., Eur. J. Appl. Physiol.* **75** (1997), 467-9.
55. Plant, D.R. *et al., Clin. Exp. Pharmacol. Physiol.* **30** (2003), 77-81.
56. Pyne, D.B. *Aust. J. Sci. Med. Sport.* **26** (1994), 49-58.
57. Radak, Z. *et al., Eur. J. Clin, Invest.* 33 (2003), 726-30.
58. Radak, Z. *et al., Free Radical Boil. Med.* **27** (1999), 69-74.
59. Ramel, A. *et al., Eur. J. Nutr.* **43** (2004), 2-6.
60. Reid, M.B. *et al., J. Appl. Physiol.* **73** (1992), 1797-804.
61. Sacheck, J.M. *et al., Free Radical Biol. Med.* **34** (2003), 1575-88.
62. Saugen, E and Vollestad, N.K. *J. Appl. Physiol.* **79** (1995), 2043-9.
63. Schippinger, G. *et al., Eur. J. Clin. Invest.* **32** (2002), 686-92.
64. Schroder, H. *et al., Eur. J.Nutr.* **40** (2001), 178-84.
65. Sen, C.K. *J. Appl. Physiol.* **79** (1995), 675-686.
66. Sen, C.K. *Sports Med. (Auckland NZ).* **31** (2001), 891-908.
67. Siems, W.G. *et al., Q. J. Med.* **92** (1999), 193-8.
68. Sjogaard, G. *et al., Eur. J. Appl. Physiol.* **57** (1988), 327-35.
69. Szczesniak, L. *et al., J Physiol Pharmacol.* **49** (1998), 421-32.
70. Terblanche, S.E. *Cell Biol. Int.* **23** (2000), 749-53.
71. Thompson, D. *et al., J. Sports. Sci.* **17** (1999), 387-95.
72. Trounce, I. *et al., Lancet.* **8** (1989), 637-639.
73. Tsai, K. *et al., Free Radical Biol. Med.* **31** (2001), 1465-72.
74. Turgut, G. *et al., Acta physiologica et pharmacologica Bulgarica.* **27** (2003), 43-5.
75. Varraso, R. *et al., Toxicol Ind Health.* **18** (2002), 269-78.
76. Vasankari, T. *et al., Clinica chimica acta; Int. J. Clin. Chem.* **234** (1995), 63-9.
77. Vincent, K.R. *et al., Eur. J. Appl. Physiol.* **87** (2002), 416-23.
78. Xu, W. *et al., Neurosci.* **126** (2004), 313-23.
79. Yu, J.G. *et al., Histochemical Cell Biol.,* **119** (2003), 383-393.

CHAPTER 6

OXIDATIVE STRESS AND ANTIOXIDANT DEFENSE: EFFECTS OF AGING AND EXERCISE

Li Li Ji
University of Wisconsin, Madison, WI

6.1 Introduction

Among the various theories attempting to explain the aging process, the free radical theory of aging has received increased recognition over the past four decades (Harman, 1956; Sohal and Weindruch, 1996). A basic tenet of this theory is that reactive oxygen species (ROS) are produced as a normal byproduct of aerobic life and that accumulation of oxidative damage caused by ROS underlies the fundamental changes found in senescence. At least three lines of evidence support the theory. (a) Aging has been shown to correlate with the production of ROS and the capacity of cellular antioxidant defense systems (Harman, 1956; Ames *et al.*, 1993; Yu, 1994). (b) An increasing number of age-related and degenerative diseases have been found to have an etiological component associated with ROS generation (Ames *et al.*, 1993) and (c) strategies that are effective to reduce oxidative stress are also found to affect aging. A clear example is dietary or caloric restriction in rodents (Sohal and Weindruch, 1996).

Research over the past two decades has shown that ROS generation is a major cause of cell and tissue injury associated with rigorous physical exertion (Jenkins, 1993; Meydani and Evans, 1993). ROS resulting either from increased oxygen consumption or from specific pathways activated during or after exercise can elicit a series of biochemical modifications to the various cellular components causing a more oxidized environment within the cell generally termed "oxidative stress". However, physical exercise is an intimate part of the life cycle as organisms need the mobility to pursue food, escape predators, and ensure reproduction. The most prominent biological change occurring during exercise is the increased metabolic rate, matched by an enhanced rate of mitochondrial respiration and oxidative phosphorylation. It is estimated that during maximal muscular contraction in men oxygen consumption at the local muscle fibers can reach as high as 100 fold of the resting levels, while the whole body oxygen consumption increases by ~20 fold (Jenkins, 1993). Such a high rate of oxygen flux may provoke increased electron "leakage" to molecular oxygen to form superoxide radicals ($O_2^{•-}$), above those found at the resting condition. Thus exercise imposes an oxidative stress in the vicinity of the mitochondria and other critical organelles essential for cell life (Yu, 1994).

In this chapter we choose to focus on the skeletal muscle for the following three reasons. (a) The general topic of aging and oxidative stress has been reviewed previously by many experts in the field. (b) Deterioration of skeletal muscle function is an important issue in medical gerontology because of the critical role of muscle for mobility and normal life. (c) Skeletal muscle has displayed some unique characteristics during aging both in terms of free radical production and antioxidant systems. It is important to keep in mind that the extent of cell oxidative damage is determined by the rate of not only ROS production, but also ROS removal provided by the antioxidant defense systems (including the capacity to repair the damage). Thus, age-related changes in muscle antioxidant capacity and possible influences of physical exercise will be emphasized.

6.2 ROS Generation with Aging and Exercise

The free radical theory of aging has been supported by strong evidence that senescent organisms produce ROS at a higher rate than their young counterparts (Harman, 1956; Sohal and Weindrch, 1996; Ames *et al.*, 1993). There are several well-documented sources of ROS production in the cell, among which the mitochondria have received particular attention because of their central role in oxidative phosphorylation (Shigenaga *et al.*, 1994). The electron transport chain (ETC) consumes ~85% of all the O_2 utilized in the cell. About 1–5% of the oxygen passing through the electron transport chain becomes various ROS as byproducts (Chance, 1979). Concentration of $O_2^{\cdot-}$ in the mitochondrial inner membrane is estimated to be $8x10^{-12}$ M (Yu, 1994), whereas the steady state H_2O_2 production of the heart mitochondria has been estimated to be 0.3–0.6 nmol/min/mg protein (Chance, 1979). Aging apparently increases ROS production in the mitochondria (Nohl and Hegner, 1978). The main site of $O_2^{\cdot-}$ and H_2O_2 production is located between complexes I and III (Sohal and Sohal, 1991). Using 2'7'-dichlorofluorescin (DCFH) as a probe, we have previously reported a 57 and 29% of increase in mitochondrial ROS production in the quadriceps muscle and heart, respectively, comparing 24 vs. 8 month old rats (Bejma and Ji, 1999; Bejma *et al.*, 2000).

The reason for the enhanced ROS production as organisms get older is not entirely clear, although several scenarios have been postulated. A major paradigm focuses on the finding that aged mitochondria have a lower level of cytochrome aa_3 content in proportion to other electron carriers, posing a potential danger that electrons may be transferred out of sequence thereby forming $O_2^{\cdot-}$ at the inner membrane (Nohl, 1986). Another hypothesis was based on the age-related increase in the hydrophilic property of the mitochondrial inner membrane wherein polyunsaturated fatty acids undergo an increasing rate of peroxidative modification with age (Chen and Yu, 1996). Pamplona *et al.,* (1999) reported that the level of unsaturated fatty acid was lower in heart mitochondria from long-lived pigeon (maximal life span ~35 years) than from rat (maximal life span 4 years), although the former has a similar body size and a higher metabolic rate. However, several recent studies indicate that Complex I is probably more important in terms of age-related changes because (a) ROS production rate is much higher with pyruvate-malate vs. succinate as respiration substrates; (b) application of Complex I inhibitor rotenone increases ROS production with Complex I substrates; and (c) antimycin A (Complex III inhibitor) increases ROS generation with pyruvate-malate, but not with succinate (Herrero and Barja, 1997; Barja and Herrero. 1998).

In addition to mitochondria, several other cellular sources of ROS production have been identified, including (a) microsomes and cytochrome P450 complex that have a special detoxification function against xenobiotics; (b) peroxisomes which produce H_2O_2 as a byproduct of amino acid oxidation; (c) cytosolic and endothelial xanthine oxidase (XO) that catalyzes the formation of $O_2^{\cdot -}$ and H_2O_2 from hypoxanthine; and (d) activated neutrophils that infiltrate injured muscle during inflammatory responses. The first two pathways are of relatively lesser importance in skeletal muscle, which contains low microsome and peroxisome contents. The latter two pathways, however, provide important cellular mechanisms for ROS generation under specific physiological and pathological conditions (see below). Integrated cellular sources of ROS generation in skeletal muscle are illustrated in Fig. 6.1.

Experimental data reveal that strenuous exercise indeed elicits an increased ROS production in aerobic tissues. Using electron paramagnetic resonance (EPR) spectroscopy, Davies et al., (1982) demonstrated that free radical signals were intensified in rat hindlimb muscle and liver after an acute bout of exhaustive running. The main free radical species was identified as semiquinone, consistent with the mitochondrial theory mentioned above. Jackson et al., (1985) reported a 70% increase in the EPR signals in electrically stimulated contracting muscle compared to the resting controls. Reid et al., (1992) demonstrated that the oxidation of DCFH was increased in isolated contracting diaphragm muscle. O'Neill et al. (1996) showed in perfused feline muscle that $^{\cdot}OH$ radical production was significantly increased during and after stimulated contraction. Using a modified DCFH assay, Bejma and Ji (1999) found that rat deep vastus laterilas (DVL) muscle immediately excised and homogenized after an exhaustive bout of running produced ROS at a higher rate than the rested control muscles. Furthermore, exercise provoked a greater effect on old muscle than young muscle working at a comparable intensity.

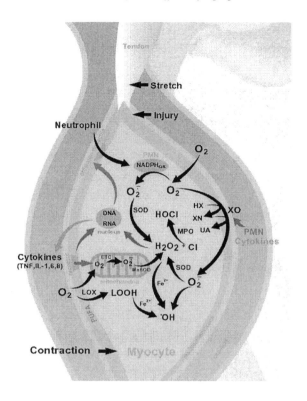

Fig. 6.1. Sources of reactive oxygen species (ROS) in muscle cells during exercise and contraction-mediated injury. ETC, election transport chain; HOCl, hypochorous acid; HX, hypoxanthine; IL, interleukine; LOX, lipooxygenase; MPO, melyeoperoxidase; PMN, polymorphoneutrophil; SOD, superoxide dismutase; TNF-α, tumor necrosis factor-α; UA, uric acid; XN, xanthine; XO, xanthine oxidase.

The above research findings lead us to believe that enhanced free radical generation at the cellular and tissue levels during physical work is a normal part of life in most animal species. Since the majority of animal species (laboratory rats and humans living in the industrialized countries are exceptions) live a fairly active life, it is not unreasonable to hypothesize that the enhanced ROS production during physical activity plays an important part in aging. This hypothesis seems to be consistent with the "rate of living" theory which postulates that the maximal life span of a species is inversely related to its metabolic rate, i.e. oxygen consumption (VO_2) per unit of body weight. Thus, smaller mammals having a higher metabolic rate due to a greater body surface area:body weight ratio and hence heat loss would have a shorter life span, whereas larger animals live relatively longer because of the lower VO₂ and possibly less ROS exposure (Sohal and Weindruch, 1996; Matsuo, 1993).

Since strenuous exercise greatly increases mitochondrial oxygen flux, it is often assumed that the majority of ROS production during exercise in aged animals is from the mitochondrial ETC. Unfortunately, the above assumption has never been proven experimentally. On the contrary, we have shown that although mitochondrial ROS production was higher in aged muscle and heart than in their young counterparts, no significant difference was observed between exercised and rested animals (Bejma and Ji, 1999; Bejma *et al.*, 2000). While it is difficult to draw conclusions with experiments using isolated mitochondria *ex vivo*, these findings do suggest that other source may be involved in contributing to ROS generation in aged animals.

6.3 ROS Generation and Inflammation

It is well-known that inflammatory cells in injured tissues can generate ROS and reactive nitrogen species (Cannon and Blumberg, 1994; Pizza *et al.*, 1998). Blood-borne polymorphoneutrophils (PMN) play a critical role in defending tissues from viral and bacterial invasion. Activation of PMN typically starts with muscle or soft tissue damage by stretching or other mechanical force (Meydani and Evans, 1993, Cannon and Blumberg, 1994). During the acute phase response, PMN penetrate to the injured tissues and release two primary factors during phagocytosis, lysozymes and $O_2^{\bullet-}$. Lysozymes facilitate the breakdown of damaged protein and cell debris by releasing proteases, whereas $O_2^{\bullet-}$ are produced by NADPH oxidase during a respiratory burst (Pyne, 1994). Cytoplasmic (CuZn) superoxide dismutase (SOD) dismutates $O_2^{\bullet-}$ to H_2O_2 that is further converted to $^{\bullet}OH$ in the presence of transition metal-ions, or to hypochlorous acid (HOCl) catalyzed by melyoperoxidase (see Fig. 6.1). While this inflammatory response is considered critical to the recovery from muscle injury, reactive oxidants released from neutrophils can also cause secondary damage (Meydani and Evans, 1993; Meydani *et al.,* 1992). For example, Best *et al.* (1999) showed in an *in situ* rabbit muscle model that stretching injury caused by maximal isokinetic contraction was associated with significant ROS generation 24 hr post treatment.   In the injured leg there was an accumulation of PMN leukocytes, accompanied by higher glutathione (GSH) levels.

Strenuous exercise can trigger releases of tumor necrosis factor-α (TNF-α), interleukin (IL)-1 and IL-6 from immune cells and/or damaged muscle tissue (Cannon and St. Pierre, 1998). During the early phase of muscle injury, these cytokines play an important role in inflammatory responses by promoting adhesion molecule expression and nitric oxide (NO) synthase induction in the endothelial cells (Gath *et al.*, 1996). The resulting increase in vasodilation due to NO production facilitates PMN and circulating cytokines to migrate to the affected area. In addition, endothelial cells from injured muscle are known to secrete TNF-α, IL-1, IL-6 and IL-8, (Pedersen *et al.*, 1998). In addition to stimulating PMN infiltration and hence ROS production, some cytokines can bind with membrane receptors and activate specific ROS-generating enzymes, such as lipooxygenase, NADPH oxidase, and XO (Flohe *et al.*, 1997). Indeed, the mechanisms involving ROS generation in contraction-induced muscle injury are very complex. Hellsten *et al.* (Hellsten *et al.*, 1997) reported in human subjects that plasma IL-1 concentration was significantly increased 90 min after an acute bout of strenuous one-leg eccentric exercise and remained elevated 4 days after exercise. Muscle XO levels were eight-fold higher in the exercised leg, which was attributed to inflammatory response measured by leukocyte invasion.

There is currently sparse data concerning whether or not aging would increase ROS generation from inflammatory cells. However, we do know that aged individuals are more susceptible to muscle injury (Brooks and Faulkner, 1994). As compared to young animals, a much smaller workload (especially eccentric) can produce mechanical injury in the aged animals. Zerba *et al.* (1990) showed that extensor digitalis longus muscle from aged mice was more susceptible to lengthening contraction than that from young mice. Treatment of SOD alleviated muscle force deficit due to eccentric injury especially in the aged animals. Thus, ROS generation and subsequent oxidative stress during muscle inflammation is considered a high risk for aged people, especially those involving in strenuous physical exertion.

6.4 Age-Related Changes in Antioxidant Systems

Aerobic organisms would not have survived during evolution if they had not developed a highly efficient and adaptive antioxidant defense system (Halliwell and Gutteridge, 1989). Antioxidants can be generally classified into two categories. The enzymatic antioxidants include SOD, catalase (CAT), and glutathione peroxidase (GPX), supported by other auxiliary enzymes such as glutathione reductase (GR), glucose 6-phosphate dehydrogenase (G6PDH) and glutathione sulfur-transferase (GST). The non-enzymatic antioxidants include antioxidant vitamins (α-tocopherol, ascorbic acid and β-carotene), thiols (mainly GSH), and a variety of low molecular weight compounds such as lipoic acid, uric acid, and ubiquinone. Each of the antioxidants is located in specific cellular sites and specialized in removing certain ROS, although considerable overlap and cooperation are demonstrated between antioxidants (Halliwell and Gutteridge, 1989). An important feature of antioxidants is that their cellular concentrations are heavily influenced by nutritional factors. Some are mandatory in the diet and others require special amino acids or trace elements for biosynthesis (Yu, 1994; Harris, 1992). The specific function and regulatory mechanism of each individual antioxidant have been studied thoroughly. However, several critical questions still remain unclear. For example, biological factors contributing to the tissue-specific changes in these enzymatic and non-enzymatic antioxidants during aging and/or in response to exercise are poorly understood. Furthermore, it is still a matter of debate whether exogenous antioxidants should be supplemented in aged and/or physically active individuals.

Profound changes in antioxidant content and activity have been observed in various tissues and organs across all species (Matsuo, 1993). According to the free radical theory of aging one might expect to see a general decline of cellular antioxidant defense capacity at old age. Available data (mostly from rodents) suggest that these changes are not uniform. Two possible mechanisms may explain this paradox. First, aging is associated with a deterioration of protein synthesis and cell differentiation capacity in most tissues, particularly in the postmitotic tissues such as heart and eye lens. Therefore, antioxidant consumption and degradation probably are not adequately replenished at old age. Second, antioxidants have demonstrated considerable adaptability in response to pro-oxidant exposure. Localized oxidative stress in specific organs, tissues, and organelles may stimulate cellular uptake and synthesis of certain antioxidants under complicated genetic, hormonal, and nutritional regulation (Harris, 1992). A well-documented example is skeletal muscle which exhibits marked increases in antioxidant enzyme activity with aging (Ji *et al.*, 1990; Lawler *et al.*, 1993; Leewenburgh *et al.*, 1994; Luhtala *et al.*, 1994; Ji *et al.*, 1990) showed that activities of all major antioxidant enzymes, such as SOD, CAT and GPX, as well as GST, GR and G6PDH, were significantly higher in the DVL (a muscle actively recruited during endurance exercise) of old vs. young rats. These changes occurred despite a general age-related decline of mitochondrial oxidative capacity. The work of Leeuwenburgh *et al.* (1994) confirmed these findings in both DVL and soleus muscles in rats. In addition, γ-glutamyl transpeptidase (GGT) activity was found to be significantly elevated indicating aged muscle might have a greater potential to take up GSH. Age-adaptation of antioxidant enzymes appears to be muscle fiber specific, with the most prominent increases found in type 1 (slow-twitch oxidative) muscles such as soleus, followed by type 2a (fast-twitch oxidative) muscles such as DVL, whereas type 2b muscles showed little effect (Leeuwenburgh *et al.*, 1994; Oh-Ishi *et al., 1995*). Luhtala *et al.,* (1994) reported that elevation of muscle antioxidant enzymes during aging was markedly affected by caloric restriction in rats. The progressive increases in CAT and GPX activities from 11 to 34 months of age were prevented by a 30% reduction of food intake, while an age-related increase in Mn SOD was also attenuated.

The mechanism responsible for increased antioxidant enzyme activities in aging skeletal muscle is still elusive. One possibility is that mitochondria from aged muscles produce more ROS that may stimulate antioxidant enzyme gene expression. This scenario is consistent with the finding that mitochondrial fractions of antioxidant enzyme activity showed a greater increase than those in the cell as a whole in the senescent skeletal muscle (Ji *et al.*, 1990; Luhtala *et al.*, 1994). Aged muscles are more susceptible to injury, which can provoke activation of the immune system causing further ROS production (Jenkins, 1993). Thus, aged muscles may be in a chronic imflammatory state wherein steady-state ROS production is elevated (Cannon and Blumberg, 1994). However, Oh-ishi *et al.* (1995) investigate gene regulation of antioxidant enzymes in the aging muscle and reported no significant change in the relative abundance of mRNA for either MnSOD or CuZnSOD in soleus muscle, although 24 month old rats had higher activities of CuZn SOD, Mn SOD, GPX and CAT than 4 month old rats. CuZn SOD protein content was also elevated in the aged muscle. Data from our laboratory also reveal no significant age differences in mRNA levels for CuZn SOD, Mn SOD or GPX in DVL muscle at old age despite prominent increases in their activities (Hollander *et al.*, 2000). Thus, gene regulation of muscle antioxidant enzymes in old age may be subjected to complicated regulation involving both transcriptional and post-transcriptional mechanisms.

Aging is associated with a decline of cellular thiol reserve in most tissues (Matsuo, 1993). However, data from our laboratory suggest that skeletal muscle and heart may be spared of this effect. Leeuwenburgh *et al.* (1994) showed that aging caused no significant alteration of GSH content or GSH/glutathione disulfide (GSSG) ratio in rat DVL muscle, whereas in soleus there was a 37% increase in GSH content in old rats along with a higher GSH/GSSG ratio. Fiebig *et al.* (1996) showed a significant increase in total glutathione content (GSH+GSSG) in the heart of 27 vs. 5 month old rats. The elevated myocardial GSH content was coupled with a two-fold increase in GGT activity, suggesting a greater potential of the γ-glutamyl cycle. Age-related alterations of antioxidant vitamins and other endogenous antioxidant systems are beyond the scope of this chapter and can be found in several excellent reviews including those cited earlier (Ames *et al.*, 1993; Yu, 1994; Shigenaga *et al.*, 1994; Matsuo, 1993).

6.5 Antioxidant Response and Adaptation to Exercise

An acute bout of strenuous exercise has been shown to alter cellular and tissue antioxidant status profoundly (Ji and Leichtweis, 1997). This is primarily caused by the high levels of ROS production during exercise, but other factors such as altered blood flow, energy status, and availability of reducing power may also influence antioxidant function in the various tissues (Ji, 1995). A good example is the effect of acute exercise on GSH homeostasis (Fig. 6.2). As the most abundant non-protein thiol source, GSH concentration in the cell is remarkably high (in the millimolar range). GSH not only serves as a substrate for GPX and GST to reduce hydrogen- and organic-peroxides, but also scavenges singlet oxygen and $\cdot$OH, and reduces tocopherol radicals and semidehydroascorbate radical, thereby preventing lipid peroxidation (Meiser and Anderson, 1983). GSH can be recycled by GR using NADPH as the cellular reducing power. Liver is the primary organ for *de novo* GSH synthesis and supplies 90% of circulatory GSH to maintain a whole-body homeostasis. Liver GSH content has been shown to decline dramatically (30 to 50%) after an acute bout of prolonged exercise (Lew *et al.*, 1985; Leeuwenburgh and Ji, 1996). This decrease is caused by three possible factors. First, hepatic GSH is oxidized to GSSG by oxygen free radicals and locally produced oxidants at a higher rate than the GR-catalyzed reduction rate. Some of the GSSG formed in the hepatocytes may be exported to the blood causing a net loss of GSH (Sies and Graf, 1985). Second, hepatic synthesis of GSH is limited by γ-glutamylcysteine synthetase (GCS) activity and requires ATP and cysteine. Strenuous exercise may reduce the availability of these compounds in the liver due to competition of other metabolic pathways. Third, and probably most important, there is an enhanced GSH output into the plasma stimulated by glucagon, catecholamines, and vasopression, which all increase their releases during prolonged exercise. Extrahepatic tissues such as kidney, heart, skeletal muscle, and erythrocytes presumably take up circulatory GSH via the γ-glutamyl cycle during exercise to ensure adequate GSH supply (Meister and Anderson, 1983). However, tissue GSH uptake requires translocation of its ingredient amino acids across the cell membranes which is limited by

GGT, an enzyme having low activity in heart and skeletal muscle (Leeuwenburgh and Ji, 1996; Leeuwenburgh *et al.*, 1996). Thus, during an acute exercise bout of relatively short duration (up to ~1 hour), GSH oxidation to GSSG in these tissues may be matched by both GR-catalyzed redox cycle and by GSH uptake from the circulation. Therefore, no substantial disturbance of GSH status is observed even at a high workload (Ji and Fu, 1992; Ji *et al.*, 1993). However, during prolonged exercise as liver GSH reserve and output diminish, tissue GSH uptake can no longer sustain the enhanced GSH oxidation by ROS, resulting in a decrease of GSH: GSSG ratio and a net GSH deficit (Lew *et al.*, 1985; Leeuwenburgh and Ji, 1996; Sen *et al.*, 1992).

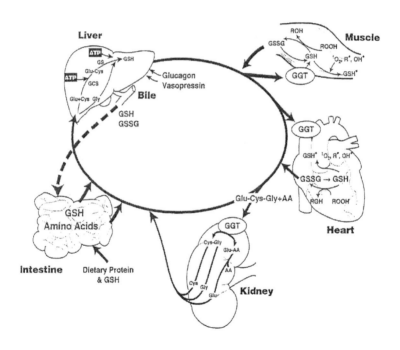

Fig. 6.2. Glutathione homeostasis in the body. GCS, γ-glutamylcystein synthetase. GGT, γ-glutamyl transpeptidase. GS, glutathione synthetase. GSSG, glutathione disulfide. ROH, alcohol. ROOH, organic hydroperoxide. R·, alkyl free radicals.

It has long been recognized that in mammals and birds antioxidant enzyme activity in skeletal muscle is higher in the wild species compared to their domestic counterparts (Burge and Neil, 1916). Within the body, tissues that have a higher metabolic rate, such as liver, heart, and brain, or those chronically exposed to ROS, such as eye lens, have greater antioxidant protection than those with lower oxygen or ROS exposure (Halliwell and Gutteridge, 1989). Antioxidant enzyme activity and GSH content vary widely in different skeletal muscle fibers. The more aerobic type 1 muscle possesses greater antioxidant potential than type 2a and type 2b muscle (Ji *et al.,* 1993). While these inter-species and inter-organ/tissue differences in antioxidant potential may reflect long-term adaptation during evolution and development, relatively short periods of oxidative stress have also been shown to induce certain antioxidants (Halliwell and Gutteridge, 1989; Harris, 1992). In humans, a higher antioxidant enzyme activity has been reported to correlate with maximal aerobic capacity (VO_2max) and trained athletes have greater SOD and CAT activities in skeletal muscle than untrained people (Jenkins, 1993).

Numerous studies have shown that antioxidant enzyme activities are elevated in skeletal muscle after endurance training involving repeated bouts of prolonged exercise (Ji, 1995). It is believed that the increased muscle ROS production constitutes the underlying reason for this training adaptation, although the regulatory mechanism is still poorly understood. However, a few highlights are worth mentioning. (a) Training adaptation of antioxidant enzymes is highly tissue and muscle fiber specific. In general, only the oxidative type of muscles, which are selectively recruited during endurance exercise (including diaphragm) show an increase in antioxidant activity (Laughlin *et al.,* 1990; Powers *et al.,* 1994). (b) The occurrence and magnitude of training adaptation in antioxidant enzymes depend largely on the amount of training, particularly the duration of each training session (Powers *et al.,* 1994). This is consistent with the notion that ROS production is roughly proportional to the level of oxygen consumption during exercise. (c) Among the various antioxidant enzymes, mitochondrial (Mn) SOD and

GPX have demonstrated the most consistent and prominent training adaptation. (d) In addition to antioxidant enzymes, GSH has displayed a training adaptation in selective muscle fibers (Leeuwenburgh *et al.*, 1997; Marin *et al.*, 1990). This may be explained in part by a higher GGT activity in these muscles, facilitating GSH uptake from the plasma during and after exercise (Leeuwenburgh *et al.*, 1997).

6.6 Exercise and Antioxidant Signaling

Finkel and Holbrook (2000) elegantly stated that the best strategy to enhance endogenous antioxidant levels may actually be oxidative stress itself, based on the classical physiological concept of hormesis. Hormesis is a Greek word meaning a sublethal dose of toxin can increase the tolerance of the organism to withstand higher doses of toxins. Exercise at high intensity is a form of oxidative stress due to the generation of ROS that exceeds the defense capacity in skeletal muscle (McArdle *et al.*, 2001; McArdle and Jackson, 2000). However, it has been consistently observed that individuals undergoing exercise training have high levels of antioxidant enzymes and certain non-enzymatic antioxidants in muscle and demonstrate greater resistance to exercise-induced or imposed oxidative stress (Ji, 1995; Sen, 1995). Presumably, these adaptations result from cumulative effects of repeated exercise bouts on the gene expression of antioxidant enzymes. The question arises as to how exercise could trigger cellular mechanisms to increase antioxidant defense, i.e. how an extracellular signal (oxidative stress) can induce an intracellular response and adaptation.

Mammalian cells are endowed with several signaling pathways that can be activated by oxidative stress. Those include the NF-κB, heat-shock transcriptional factor 1 (HSF-1), and P53 pathways, as well as mitogen-activated protein kinase (MAPK) and PI(3)K/Art that regulate the first three pathways through phosphorylation (Finkel and Holbrook, 2000) (Fig. 6.3). ROS have been suggested to play the role of second or third messenger in the activation of these redox-sensitive pathways (Schreck and Baeuerle, 1991). Recent evidence suggests that a single bout of muscular contraction, especially eccentric contraction, can activate MAPK pathway in human skeletal muscle (Aronson *et al.,* 1997; Boppart *et al.,* 1999; Wildegren *et al.,* 1998). Sixty minutes after an acute bout of one-leg cycling activity of MAPK-activated protein kinase 2 was increased by 300% (Krook *et al.,* 2000). Furthermore, Extracellular signal-Regulated Kinase (ERK) and p38 MAPK activity was increased in rat slow- and fast-twitch skeletal muscle after electrically stimulated contraction (Wretman *et al.,* 2000). Nader and Esser (2001) showed that immediately after an acute bout of treadmill running ERK and p38 were activated in rat soleus and tibialis muscles. Activation of various kinases involved in the MAPK pathway can lead to the sequential phosphorylation of a series of proteins, resulting in increased expression of c-Jun, a subunit of the transcription factor activator protein-1 (AP-1) (Pulyerer *et al.,* 1991). Alternatively, it may phosphorylate downstream kinases such as p90 ribosomal S6 kinase (p90rsk), which activity was found to increase up to 25 fold in human muscle after exercise (Krook *et al.,* 2000). Whether exercise-activated MAPK signaling pathway plays a role in the elevated antioxidant gene expression has not been elucidated.

Mechanisms of NFκB-induced signaling in response to oxidative stress are well defined (Meyer *et al.,* 1994; Allen and Tresini, 2000). ROS have been shown to activate several kinases that phosphorylate serine residue 19 and 23 on the inhibitory subunit (IκB) of NF-κB, causing its ubiquitination and release from the NFκB complex. The p50 and p65 dimer subsequently translocates into the nucleus and binds to the κB domain of the target gene promoter, leading to transcriptional activation. Cellular redox status influences NF-κB activation profoundly (Flohe *et al.,* 1997; Sen and Packer, 1996). Although ROS and other pro-oxidant cytokines such as TNF-α initiate IκB dissociation, binding of activated and translocated p50 and p65 dimer to DNA sequence requires a reduced cellular milieu with possible participation of GPX and thioredoxin (Meyer *et al.,* 1994).

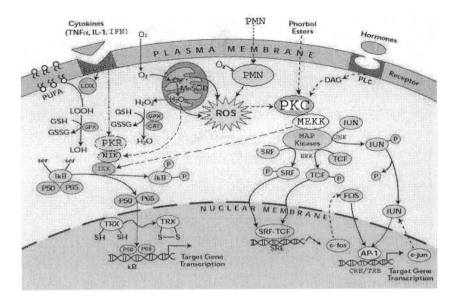

Fig. 6.3. NFκB and MAPK signaling pathway in the cell. CAT, catalase. CRE, cAMP-response element. DAG, diacylglycerol. ERK, extracellular signal-regulated kinase. GPX, glutathione peroxidase. IFN, interferin. IKK, IκB kinase. JNK, c-Jun amino-terminal kinase LOOH, lipoperoxide. LOH, hydroxylipid. LOX, lipooxygenase. MKK, MAP kinase kinase. Mn SOD, manganese superoxide dismutase. NIK, NFκB induced kinase. PLC, phospholipase C. PKC, protein kinase C. PKR, double strand-RNA activated protein kinase. PMN, polymorphoneutrophil. PUFA, polyunsaturated fatty acid. ROS, reactive oxygen species. SRE, serum response element. SRF, serum response factor. TCF, ternary complex factor. TRE, tetradecanoylphorbolacetate response element. TNFα, tumor necrosis factor α. TRX, thioredoxin.

Several antioxidant enzymes contain NFκB and AP-1binding sites in their gene promoter region, such as Mn SOD and GCS (Allen and Tresini, 2000). Therefore, they are potential targets for exercise-activated upregulation via NFκB signaling pathways. Hollander *et al.* (2001) investigated the time course after an acute bout of treadmill running on Mn SOD gene expression in rat skeletal muscle. In both type 2a (DVL) and 2b (SVL), NFκB binding was significantly increased ~2 h after the acute exercise bout and remained elevated during the following

48 hr. AP-1 binding in these two muscle types were also dramatically increased by acute exercise reaching peak at 30 min, but returned to resting levels within a few hours. mRNA abundance for MnSOD in DVL was increased in the exercised rats, whereas an increase in MnSOD protein level was observed in SVL only after 48 h. These data suggest that an acute bout of exercise may represent a sufficiently large oxidative stress to activate MnSOD gene transcription via NFκB signaling.

In order to delineate the molecular events pertaining to the ROS-induced gene expression of MnSOD shown in the above study, we recently performed two studies to investigate the effects of an acute bout of physical exercise on the NFκB signaling pathway in rat skeletal muscle (Ji *et al.*, 2004). In study 1, a group of rats ran on the treadmill at 25 m/min, 5% grade, for 1 h or until exhaustion, and compared with a second group injected with two doses of pyrrolidine dithiocarbamate (PDTC, 100 mg/kg, i.p.), a potent NFkB inhibitor, 24 and 1 hr prior to the acute exercise bout. Three additional groups of rats were injected either 8 mg/kg (i.p.) of lipopolysaccharide (LPS), or 1 mmol/kg (i.p.) *t*-butylhydroperoxide (tBHP), or saline as controls and killed at resting condition. The exercised rats showed significantly higher levels of NFκB binding and P50 protein content in the muscle nuclear extracts compared to control rats. Cytosolic IκBα and IκB kinase (IKK) contents were decreased, whereas phospho-IκB and phospho-IKK contents were increased, comparing exercised vs. control rats. The exercise-induced activation of NFκB signaling cascade was partially abolished by PDTC treatment. LPS, but not tBHP, treatment mimicked and exaggerated the effects observed in exercised rats. In Study 2, the time course of exercise-induced NFκB activation was examined. Highest levels of NFκB binding and P65 content were observed at 2 h post exercise. Decreased cytosolic IκBα and increased phosphor-IκBα content were found at 0-1 h post exercise, whereas P65 reached peak levels at 2-4 h. These data suggest that the NFκB signaling pathway can be activated in a redox-sensitive manner during muscular contraction, presumably due to increased oxidant production. The cascade of intracellular events may be the overture to elevated gene expression of MnSOD reported in our previous study (Hollander *et al.*, 2001).

6.7 Training Adaptation in Aging Skeletal Muscle

Senescent skeletal muscles lose considerable regenerative ability due to a decreased rate of protein turnover, satellite cell population and proliferative capacity (Carlson, 1995). Increased protein breakdown, partly explained by oxidative damage and consequently selective degradation, coupled with decreased protein synthesis, results in a progressive decline in oxidative enzyme levels and energy production in aging skeletal muscle (Jenkins, 1993; Hansford, 1983). Endurance training has been shown to effectively restore losses of muscle protein content and mitochondrial oxidative capacity in old age (Fitts et al., 1984). However, there are conflicting data concerning training adaptation of antioxidant systems in aged skeletal muscle. Ji et al. (1991) showed that along with increased activities of muscle citrate synthase, malate dehydrogenase and lactate dehydrogenase, training also resulted in a 60% increase in GPX activity in 27.5 month old Fischer 344 rats. Hammeren et al. (1993) reported a significant increase in GPX activity with training in several skeletal muscles of old Fischer 344 rats. However, Leeuwenburgh et al. (1994) could not find antioxidant enzyme adaptation in either DVL or soleus muscle from old rats. The failure to induce antioxidant enzyme levels in aged muscle obviously cannot be accounted for by a reduced cell proliferative capacity, because training successfully increased mitochondrial oxidative enzyme activity and protein content (Fitss et al., 1984; Ji et al., 1991; Hammeren et al., 1993) A possible explanation is that antioxidant enzyme activities in the senescent muscle are already elevated. Therefore the training threshold is raised requiring greater intensity to provoke a training effect.

What is the physiological significance of training adaptation in antioxidant enzymes in aged muscle? When exercising at a given workload, oxygen consumption is no different between trained and untrained muscle. Therefore, endurance trained aged muscle has the advantage of distributing the oxidative flux (electron flux) among increased mitochondrial ETC as compared to the untrained muscle (Davies et al., 1982). This should theoretically result in a lesser degree of ROS production. Furthermore, trained muscles have higher levels of antioxidant enzyme activity and GSH content, providing a greater ROS removal. Thus, training in old age should reduce muscle oxidative injury during acute physical exertion. Indeed, DVL and soleus muscles from the aged rats displayed significantly lower MDA levels than their sedentary counterparts after endurance training (Leeuwenburgh *et al.*, 1994). However, these data should be interpreted with caution because muscle MDA levels could also be influenced by the rate of efflux from the muscle (Halliwell and Gutteridge, 1989). The low MDA content may result from enhanced output due to circulatory training adaptation in the old animals.

The effect of endurance training on mitochondrial respiratory function and resistance to ROS has been studied in aged skeletal muscle (Hammeren *et al.*, 1993). While training elicited no apparent improvement on state 3 respiration rate and respiratory control index (RCI), the magnitude of RCI inhibition by equal doses of $O_2^{\bullet-}$ and H_2O_2 was significantly smaller in the trained vs. control rats. If these results from this *in vitro* study could be extrapolated to *in vivo* condition, aged skeletal muscle undergoing exercise training may be more capable of maintaining adequate mitochondrial function when exposed to acute oxidative stress.

There is a general concern that aged animals are subjected to a deficiency of antioxidant nutrients (Starnes *et al.*, 1989). Exercise training may potentially exacerbate this nutritional deficiency due to increased consumption or decreased dietary intake, or both. However, a clear consensus in this important area of gerontology is lacking. Starnes *et al.* (1989) compared muscle α-tocopherol levels between trained and untrained 24 month old rats and found a significant decrease in the trained muscle. Old trained rats also displayed a higher level of muscle thiobarbituric acid reactive substance (TBARS) when challenged by ascorbic acid and ferrous ion. These findings are consistent with the earlier data concerning training effect on muscle vitamin E levels in the young rats (Aikawa *et al.*, 1984; Gohil *et al.*, 1987). Taken together, it seems that aged skeletal muscle does not necessarily lose adaptability to training. It is still capable of up-regulating antioxidant enzyme synthesis and maintaining GSH status. However, aged muscle may be susceptible to deficit of antioxidant vitamins that the body can not synthesize. It is conceivable that physically active elderly individuals may benefit from dietary antioxidant supplementation.

6.8 Aging, Exercise and Phytochemical Supplementation

Nature offers an abundance of resources of antioxidants, most of which are present in fruits and vegetables known as phytochemicals (Hertog, 1996). Most of the phytochemicals are in the chemical form of phenolic compounds that have the ability of quenching or reducing ROS due to their redox properties. Tocols (such as tocopherols and tocotrienols), flavonoids (such as soy isoflavone, tea catechins and anthocyanidines), monophenolic acids (such as caffeic acid and ferulic acids) and polyphenolic acids (such as avenathramides) are the most common antioxidant phytochemicals.

We have become very interested in exploring this promising field that has great potential in providing effective protection against aging- and exercise- associated oxidative damage. Recently we conducted a study wherein antioxidant properties of North American Ginseng (*Panax Quinquefolium*) were investigated in young and old rats fed a ginseng-supplemented diet for a period of 4 months (Fu and Ji, 2003). Female Fischer 344 rats at 4 or 22 months of age were randomly divided into three groups, fed either a AIN-93G formula-based control diet, or a diet containing 0.5 g/kg (low-dose) or 2.5 g/kg (high-dose) dry ginseng power for 4 months. Oxidant generation, measured with DCFH, was significantly lowered with ginseng feeding in the homogenates of heart, soleus, and DVL muscle in both young and old rats, and the effects were dose-dependent. SOD activity was elevated in heart, DVL and soleus muscle with high and low dose of ginseng feeding, respectively. Rats fed with high-dose ginseng diet showed higher GPX activity in DVL and soleus muscle, and elevated citrate synthase activity in the heart and DVL. Further, protein carbonyl content was attenuated with high dose in the heart and with both doses in DVL. These data indicate that ginseng supplementation can prevent age-associated increase in oxidant production and oxidative protein damage in rats. These protective effects are partially explained by elevated antioxidant enzyme activities in the various tissues.

6.9 Summary

Skeletal muscle is under increasing oxidative stress during aging. This is evidenced by the higher levels of ROS generation, oxidative damage to proteins and lipids, and elevated antioxidant enzyme activity in aged muscle. Physical exercise promotes the generation of ROS in skeletal muscle primarily via the mitochondrial electron transport chain, but multiple cellular sources may contribute to the increased ROS during and after exercise under various pathophysiological conditions. Senescent muscle appears to be more susceptible to exercise-induced oxidative damage possibly due to its greater vulnerability to mechanical injury and associated inflammatory response. Exercise training provides a clear benefit by up-regulating muscle antioxidant enzymes and glutathione status, and this adaptability appears to be blunted but not entirely lost in old age. However, intensive exercise can deplete muscle antioxidant vitamin levels which may compromise the muscle's overall antioxidant protective margin. Thus, choosing appropriate mode and intensity of exercise is essential for the elderly who participate in physical activity. It is also important to ensure sufficient dietary antioxidant intake to minimize a possible oxidative stress. Finally, it is reasonable to consider dietary supplementation, either from the manufactured or natural source, to increase protection.

References

1. Aikawa, K.M. *et al., Biosci. Rep.* **4** (1984), 253-257.
2. Allen, R.G. and Tresini, M. *Free Rad. Biol. Med.* **28** (2000), 463-499.
3. Ames, B.N. *et al., Proc. Natl. Acad. Sci.* **90** (1993), 7915-7922.
4. Aronson, D. *et al., J. Clin. Invest.* **99** (1997), 1251-1257.
5. Barja, G. and Herrero, A., *J. Bioener. Biomem.* **30** (1998), 235-243.
6. Bejma, J. and Ji, L.L. *J. Appl. Physiol.* **87** (1999), 465-470.
7. Bejma, J. *et al., Acta Physiol. Scand.* **169** (2000), 343-351.
8. Best, T. *et al., J. Appl. Physiol.* **87** (1999), 74-82.

9. Boppart, M.D. *et al., J. Appl. Physiol.* **87** (1999), 1668-1673.
10. Brooks S. and Faulkner, J.A. *Med. Sci. Sports Exer.* **26** (1994), 432-439.
11. Burge, W.E. and Neil, A.J. *Am. J. Physiol.* **43** (1916-1917), 433-437.
12. Cannon, J.G. and Blumberg, J.B. *Exercise and Oxygen Toxicity*, ed. Sen, C.K., Packer, L., Hanninen, O. (Elsevier Science, New York, 1994), 447-79.
13. Cannon, J.G. and St. Pierre, B.A., *Mol. Cell. Biochem.* **179** (1998), 159-167.
14. Carlson, B.M. *J. Gerontol.* **50** (1995), 96-100.
15. Chance, B.H. *Physiol. Rev.* **59** (1979), 527-605.
16. Chen, J. J. and Yu, B.P., *Free. Rad. Biol. Med.* **17** (1996), 411-418.
17. Davies, K.J.A. *et al., Biochem. Biophys. Res. Comm.* **107** (1982), 1198-1205.
18. Fiebig, R. *et al., Age.* **19** (1996), 83-89.
19. Finkel, T. and Holbrook, N. *Nature.* **408** (2000), 239-247.
20. Fitts, R.H. *et al., Mech. Ageing Dev.* **27** (1984), 161-172
21. Flohe, L.R. *et al., Free Rad. Biol. Med.* **22** (1997), 1115-1126.
22. Flohe, L.R. *et al., Free Rad. Biol. Med.* **22** (1997), 1115-1126.
23. Fu, Y. and Ji, L.L. *J. Nutr.* **133** (2003), 3603-3609.
24. Gath, I. *et al., FASEB J.* **10** (1996), 1614-1620.
25. Gohil, K. *et al., J. Appl. Physiol.* **63** (1987), 1638-1641.
26. Halliwell, B. and Gutteridge, J.M.C. *Free Radicals in Biology and Medicine*, (2nd ed. Clarendon Press, Oxford, England, 1989).
27. Hammeren, J. *et al., Int. J. Sports Med.* **13** (1993), 412-416.
28. Hansford, R.G. Biochim. *Biophys. Acta.* **726** (1983), 41-80.
29. Harman, D. *J. Gerontol.* **11** (1956), 298-300.
30. Harris, E.D. *FASEB J.* **6** (1992), 2675-2683.
31. Hellsten, Y.U. *et al., J. Physiol.* **498** (1997), 239-248.
32. Herrero A, and Barja, G. *Mech. Ageing Dev.* **98** (1997), 95-111.
33. Hertog, M.G. *Proc Nutr Soc.* **55** (1996), 385-397.
34. Hollander, J. *et al., Mech. Ageing Dev.* **116** (2000), 33-45.
35. Hollander, J. *et al., Pflug. Arch. (Eur. J. Physiol.).* **442** (2001), 426-434.
36. Jackson, M.J. *et al., Biochim. Biophys. Acta.* **847** (1985), 185-190.
37. Jenkins, R.R. *Intl. J. Sports Nutr.* **3** (1993) 356-375.
38. Ji, L.L. and Fu, R.G. *J. Appl. Physiol.* **72** (1992), 549-554.
39. Ji, L.L. *et al., Am. J. Physiol.* **258** (1990), R918-R923.
40. Ji, L.L. *et al., FASEB J.* **18** (2004), 1499-1506,
41. Ji, L.L. *et al., Gerontol.* **37** (1991), 317-325.
42. Ji, L.L. *et al., J. Appl. Physiol.* **73** (1993), 1854-1859.
43. Ji, L.L. *Exercise and Sport Science Review.* ed. Holloszy, J. (Williams & Wilkins Co. Baltimore, Maryland, 1995), 135-166.
44. Ji, L.L. and Leichtweis, S.L. *Age.* **20** (1997), 91-106.
45. Katz, M.L. and Robinson, W.G. *Free Radicals, Aging, and Degenerative Diseases.* ed. Johnson, J.E., Walford, R., Harman, D. and Miguel, J. (Alan R. Liss, Inc., New

York, New York., 1986), 221-262.
46. Krook, A. *et al., Am. J. Physiol.* **279** (2000), R1716-21.
47. Laughlin, M.H. *et al., J. Appl. Physiol.* **68** (1990), 2337-2343.
48. Lawler, J.M. *et al., Am. J. Physiol.* **265** (1993), R1344-R1350.
49. Leeuwenburgh, C. and Ji, L.L. *J. Nutr.* **126** (1996), 1833-1843.
50. Leeuwenburgh, C. *et al. Mol.Cell Biochem.* **156** (1996), 17-24.
51. Leeuwenburgh, C. *et al., Am. J. Physiol.* **267** (1994), R439-R445.
52. Leeuwenburgh, C. *et al., Am. J. Physiol.* **272** (1997), R363-R369.
53. Lew, H. *et al., FEBS Lett.* **185** (1985), 262-266.
54. Luhtala, T.A. *et al., J. Gerontol.* **49** (1994), B321-B328.
55. Marin, E. *et al., Acta. Physiol. Hung.* **76** (1990), 71-76.
56. Matsuo, M. *Free Radicals in Aging.* ed. Yu, B.P. (CRC Press, Boca Raton, Florida, 1993), 143-181.
57. McArdle, A. and Jackson, M.J. *J. Anat.* **197** (2000), 539-541.
58. McArdle, A. *et al., Am. J. Physiol.* **280** (2001), C621-627.
59. Meister, A. and Anderson, M.E. *Ann. Rev. Biochem.* **52** (1983), 711-760.
60. Meydani, M. and Evans, W.J. *Free Radical in Aging*, ed. Yu, B.P. (CRC Press. Boca Raton, Florida, 1993), 183-204.
61. Meydani, M. *et al., Ann. N.Y. Acad. Sci.,* **669** (1992), 363-374.
62. Meyer, M. *et al., Chemico-Biol. Interact.* **91** (1994), 91-100.
63. Nader, G.A and Esser, K.A. *J. Appl. Physiol.* **90** (2001), 1936-42.
64. Nohl, H. *Free Radicals, Aging, and Degenerative Diseases.* ed. Johnson, J.E. (Alan R.Liss, Inc. New York, New York, 1986), 77-98.
65. Nohl, H. and Hegner, D. *Eur. J. Biochem.* **82** (1978), 563-547.
66. O'Neill, C.A. *et al., J. Appl. Physiol.* **81(3)** (1996), 1197-1206.
67. Oh-Ishi, S. *et al., Mech. Age. Dev.* **84** (1995), 65-76.
68. Pamplona, R.M. *et al., Mech Ageing Dev.* **106** (1999), 283-296.
69. Pedersen, B.K. *et al., Can. J. Physiol Pharmacol.* **76** (1998), 505-511.
70. Pizza, F.X. *et al., J. Leukoc. Biol.* **64** (1998), 427-433.
71. Powers, S.K. *et al., Am J. Physiol.* **266** (1994), R375-R380.
72. Pulverer, B.J. *et al., Nature.* **353** (1991), 670-674.
73. Pyne, D.B. *Sports Med.,* **17** (1994), 245-258.
74. Reid, M.B. *et al., J. Appl. Physiol.* **73** (1992), 1797-1804.
75. Schreck, R. and Baeuerle, P.A. *Trends Cell Biol.* **1** (1991), 39-42.
76. Sen, C.K. *J. Appl. Physiol.* **79** (1995), 675-686.
77. Sen, C.K. and Packer, L. *FASEB J.* **10** (1996), 709-720.
78. Sen, C.K. *et al., J. Appl. Physiol.* **73** (1992), 1265-1272.
79. Shigenaga, M.K. *et al., Proc. Natl. Acad. Sci. USA.* **91** (1994) 10771-10778.
80. Sies, H. and Graf, P. *Biochem. J.* **226** (1985), 545-549.
81. Sohal , R.S. and Sohal, B.H. *Mech. Age. Dev.* **57** (1991), 187-202.
82. Sohal, R.S. and Weindruch, R. *Science.* **273** (1996), 59-63.
83. Starnes, J.W. *et al., J. Appl. Physiol.* **67** (1989), 69-75.

84. Widegren, U. *et al., FASEB J.* **12** (1998), 1379-1389.
85. Wretman , C. *et al., Acta Physiol Scand.* **170** (2000), 45-99
86. Yu, B.P. *Physiol. Rev.* **74** (1994), 139-162.
87. Zerba, E. *et al., Am. J. Physiol.* **258** (1990), C429-C435.

CHAPTER 7

MUSCLE, OXIDATIVE STRESS, AND AGING

J.S. Moylan[1], W.J. Durham[2], and M.B. Reid[1]

[1]*University of Kentucky, Lexington, KY*
[2]*Baylor College of Medicine, Houston, TX*

7.1 Introduction

As we age, we lose muscle mass and the muscle we retain is weaker. In addition, the remaining muscle is more susceptible to and recovers more slowly from damage. This progressive loss in mass, strength and adaptive ability is called sarcopenia. Many factors are implicated in sarcopenia. The most apparent causes being lack of use, the depletion of stem cells for muscle regeneration and a decline in anabolic hormones such as growth hormone and sexual steroids. However, a growing body of evidence indicates that the normal generation of free radicals by skeletal muscle plays a major role in age-related muscle loss.

Since the early 1980s, biologists have recognized that skeletal muscle generates free radicals. Of particular interest are two closely related reduction-oxidation (redox) cascades: reactive oxygen species (ROS) and nitric oxide (NO·) derivatives. ROS and NO· derivatives are continually synthesized by skeletal muscle fibers via processes that tick along slowly under resting conditions and markedly accelerate during strenuous exercise. ROS and NO· rapidly oxidize other biological molecules including DNA, protein and lipids. This oxidation at low levels can be part of normal cellular function. However, high levels can cause cellular damage. The biological activities of muscle-derived ROS and NO· are buffered by antioxidant systems within skeletal muscle cells. ROS and NO· derivatives that escape buffering have an instantaneous effect on force production and muscle endurance. They also exert a delayed effect by acting as signaling molecules that influence muscle gene expression.

Recent studies provide evidence that aging exaggerates the effects of muscle-derived oxidants. This increased oxidant sensitivity may reflect the effects of accelerated ROS production, a decreased efficiency in the adaptive response to oxidative stress and an increased sensitivity of signaling molecules to oxidative modification. As a result, redox-sensitive processes are altered, including contractile function and gene expression. These alterations appear to contribute to the reduced muscle mass and strength of aged individuals.

7.2 Redox Balance and Aging: ROS and NO·

ROS refers to the cascade of oxygen derivatives generated by skeletal muscle. Under resting conditions generation of ROS is relatively slow, while during exercise rates accelerate (Reid, 2001). Typically, the parent molecule in this cascade is the superoxide anion ($\cdot O_2^-$), a free radical generated by addition of an electron to the outer orbital of oxygen (O_2). Superoxide anions undergo electron exchange reactions to form derivatives that include singlet oxygen ($^1O_2$), hydrogen peroxide (H_2O_2) and hydroxyl radicals ($\cdot OH$). All of these derivatives retain redox activity to a greater or lesser degree and each can influence redox-sensitive processes within the cell. In most eukaryotic cells, the principle source of superoxide anions is the mitochondrial electron transport chain. Unpaired electrons are lost at several steps in the chain. Ubiquinone (CoQ, Fig. 7.1) is the primary site of this electron "leak" and oxygen functions as the electron acceptor.

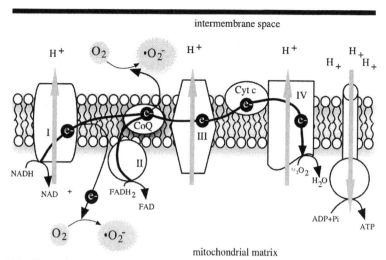

Fig. 7.1. Sites of superoxide production in the mitochondrial electron transport chain. Diagram of the mitochondrial electron transport chain showing Ubiquinone (CoQ) and Complex I as the main sources of electrons involved in superoxide generation.

The formation of ROS accounts for 2–3% of mitochondrial oxygen consumption. Superoxide anions can also be generated by enzymes such as NAD(P)H oxidases, cyclooxygenases, lipoxygenases and NO synthases. The biological importance of these production sites has not been established for skeletal muscle (Reid, 2001).

ROS production by muscle accelerates with age (Fielding and Meydani, 1997). This increase is attributed to an age-dependent decline in mitochondrial function characterized by slowed turnover, decreased efficiency, and increased electron displacement from the transport chain (Navarro *et al.*, 2001; Pesce, Cormio *et al.*, 2001; Wallace, 2001). It has been hypothesized that this decline in function is a consequence of mitochondrial ROS production. Since mitochondria are a primary ROS source, they may also be especially susceptible to oxidative damage and subsequent mitochondrial dysfunction. The "Mitochondrial Theory of Aging" describes this phenomenon as a vicious cycle where mitochondrial production of ROS leads to oxidative damage, mitochondrial dysfunction and increased ROS production (Harman,

1972). In addition, both collectively and individually these actions lead to poor muscle function. Excess ROS causes contractile dysfunction, regulatory changes and muscle protein degradation. Oxidative damage of muscle proteins may also lead to contractile dysfunction and muscle protein degradation. Finally, mitochondrial dysfunction leads to inefficient energy production and compromised muscle function (Fig. 7.2).

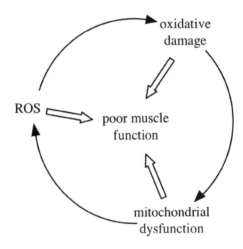

Fig. 7.2. Mitochondrial Theory of Aging. This "vicious cycle" depicts mitochondrial production of ROS causing oxidative damage that leads to mitochondrial dysfunction and increased ROS production. Each component in the cycle compromises muscle function.

NO is generated by the enzyme NO synthase (NOS), which metabolizes L-arginine to NO. Skeletal muscle fibers continually produce NO from two NOS isoforms, the type I or neuronal isoform (nNOS) and the type II or endothelial isoform (eNOS) (Reid, 2001). NO production increases with exercise and septic shock (Alvarez and Boveris, 2004; Patwell, *et al.*, 2004) and has been implicated as a mediator of muscle weakness and loss with chronic disease and disuse. Analogous to ROS, NO is the parent molecule in a cascade of NO derivatives that retain varying degrees of biological activity. ROS and NO can directly interact. For example, NO reacts with the superoxide anion to form peroxynitrite, a highly unstable and potentially damaging molecule.

Despite similarities with the ROS cascade, aging appears to have the opposite effect on NO metabolism. Both NOS expression and NO production decline with age (Richmonds *et al.,* 1999). These changes predict NO signaling and NO-mediated processes have only a minor role in the oxidant induced muscle dysfunction of aged individuals. However, other studies indicate NO may contribute to contractile dysfunction. Several researchers have shown that NO modifications accumulate in selective proteins of aged muscle. NO modifies proteins through an interaction with tyrosine residues to yield nitrotyrosine. In aged rats, the number of nitrotyrosines found in the type 2 sarcoplasmic reticulum Ca-ATPase (SERCA2a) is increased. This nitrosylation is associated with significant inactivation of SERCA2a Ca^{2+} uptake (Viner *et al.,* 1999). This inactivation could contribute to contractile dysfunction through changes in Ca^{2+} regulation. In addition, the metabolic enzymes, β-enolase and α-fructose aldolase preferentially accumulate nitrotyrosine residues with age (Kanski *et al.,* 2003). It is not known whether the modified enzymes have altered function. However, these modifications may impair energy metabolism in aging muscle.

7.3 Antioxidants

Antioxidants and stress-related proteins within muscle cells act to limit the biological effects of ROS and NO (Powers and Hamilton, 1999). Reduced glutathione (GSH) is the most plentiful antioxidant in muscle. Present in millimolar concentrations, it is the most important antioxidant biologically. GSH inactivates a variety of oxidant species directly and buffers peroxides via the enzymatic reaction catalyzed by glutathione peroxidase. Both pathways yield the oxidized glutathione dimmer (GSSG), a by-product that is recycled back to GSH via the enzyme glutathione reductase. The balance of this dynamic process is reflected by the ratio of GSH to GSSG. This ratio is a standard marker of cellular redox status. In addition to glutathione-mediated buffering, muscle cells contain the antioxidant enzymes superoxide dismutase (SOD) and catalase. SOD dismutates superoxide anions to hydrogen peroxide and catalase dehydrates hydrogen peroxide to water. Antioxidant nutrients are also important in managing muscle cell redox status. These nutrients include α-tocopherol, vitamin C, carotenoids, β-carotene and α-lipoic acid.

Antioxidant activity in skeletal muscle of aged individuals is not obviously compromised. Vitamin C levels decline, but GSH regulation and SOD and catalase activity are maintained or even increase (Pansarasa *et al.*, 1999; Pansarasa *et al.*, 2000). Similarly, vitamin E levels can equal or exceed the levels in normal adult muscle. Although antioxidant levels appear to be maintained or increase, it is possible these levels are inadequate to buffer the rise in oxidant production that occurs with aging. In support of this idea, studies have shown that indicators of oxidative damage increase with age. These include urinary 8-OHdG, an indicator of DNA damage, and protein carbonyls (indicator of protein oxidation) accumulate with age (Gianni *et al.*, 2004).

In addition to antioxidant systems, stress response proteins, namely heat shock proteins (HSPs), also oppose oxidative damage (McArdle *et al.*, 2000). HSPs generally serve as chaperones for newly synthesized proteins. They minimize protein aggregation and aid in proper folding and localization. HSP levels are increased in skeletal muscle following the oxidative stress that accompanies strenuous exercise. However, there is evidence that this response is severely blunted with age (Spiers *et al.*, 2000; Vasilaki *et al.*, 2002). This blunted response may be partially responsible for the increased susceptibility of aged muscle to damage and loss.

7.4 Aging, Redox Signaling, Muscle Function and Contractile Function

Aged muscle produces less force during contractions. This may in part be due to altered redox balance. Under normal conditions, muscle redox status continually modulates force production in skeletal muscle (Reid, 2001). When redox status is optimal, force production is maximal. When redox status is shifted toward an oxidizing state, ROS and NO are elevated and force is depressed. Consistent with this paradigm, ROS is elevated in aged muscle. In addition, aged muscle shows increased products of protein, lipid and DNA oxidation, and increased expression of ROS induced messages (Bejma and Ji, 1999; Zainal *et al.*, 2000; Kayo *et al.*, 2001). In contrast, both NOS expression and NO production are decreased in aged individuals. Therefore, NO may have a limited role in age-related contractile dysfunction.

A model for how altered redox status mediates contractile dysfunction in aged muscle is shown in Fig. 7.3. In resting healthy adult muscle, moderate ROS levels are required for basal force production. A moderate increase in ROS, such as that produced with moderate exercise is required for maximum force production. While excess ROS, such as that produced with strenuous exercise, inhibits force. Figure 7.3 shows this effect as a bell shaped continuum of contractile function over a range of redox states. It has been proposed that resting aged muscles have ROS levels similar to those produced by young muscle under strenuous exercise. Therefore, in resting aged muscle, contractile function is suboptimal and even a moderate increase in oxidants produces a rapid decline in function (Reid *et al.*, 1993). This proposal is consistent with the observations that aged muscle have increased oxidant production and increased sensitivity to oxidative stress (Lawler *et al.*, 1997; Ji *et al.*, 1998; Bejma and Ji, 1999; Richmonds and Kaminski, 2000).

ROS are thought to depress skeletal muscle force by altering the function of one or more redox-sensitive contractile regulatory proteins. Two opposing theories address this phenomenon. Studies in intact muscle fibers have shown that exogenous ROS can markedly depress force without altering calcium regulation (Andrade *et al.*, 1998; Andrade *et al.*, 2001). These findings argue that the proteins most sensitive to oxidative modification are located downstream of the calcium signal, for example, actin, myosin, troponin or other myofibrillar proteins (Reid, 2001). An alternate body of research implicates a disruption in Ca^{2+} regulation that leads to contractile dysfunction. Data from aged rats indicate that the Ca^{2+}-dependent ATPase, SERCA2a, exhibits marked dysfunction that is linked to extensive oxidative modification

(Schoneich *et al.*, 1999). The molecule contains fewer reduced cysteines and increased nitrotyrosine residues. These observations suggest that peroxynitrite contributes to oxidative damage that leads to deregulation of Ca^{2+} flow and subsequent impaired contractile function in aged muscle.

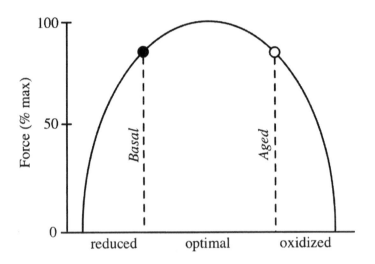

Fig. 7.3. Redox sensitivity of muscle force production. *Curve* depicts biphasic response of force as muscle redox status is altered. *Closed circle* is proposed status of healthy adult muscle under basal conditions. *Open circle is* proposed status of aged muscle. Model adapted from Reid *et al.* (1993).

Finally, oxidation of membrane lipids could cause membrane changes that affect Ca^{2+} movement. This idea is supported by the observation that muscle from patients with chronic fatigue syndrome is under oxidative stress. Muscle from these patients has altered membrane fluidity and fatty acid composition (Fulle *et al.*, 2000; Fulle *et al.*, 2003).

In addition to effects on muscle function, increased ROS may affect muscle regeneration. Following damage, muscle is replenished from a reservoir of satellite cells that have the ability to proliferate and fuse to form new muscle. These cells have a limited lifespan. When placed into a tissue culture environment, satellite cells from aged individuals go through fewer population doublings or reach senescence sooner than those from young individuals (Renault *et al.*, 2002). Senescence in human fibroblasts can be prematurely induced by oxidative stress (von Zglinicki *et al.*, 1995; Dumont *et al.*, 2001). It has been hypothesized that oxidative stress also affects the proliferative capacity of satellite cells. In support of this hypothesis, Renault *et al.* (2002), treated quadriceps myoblasts with H_2O_2 for 30 minutes. These myoblasts retained the ability to fuse and form muscle but their lifespan in culture was reduced by 10%.

7.5 Gene Expression

The age-related increase in ROS is thought to affect complex signaling mechanisms that lead to altered gene expression (Sen and Packer, 1996). In healthy adults, ROS and NO· promote expression of genes that facilitate the adaptation of skeletal muscle to stress or exercise. At least 2 major redox sensitive systems are involved. By far the most extensively studied response is the capacity of oxidative stimuli to induce cytoprotective proteins such as antioxidants and HSPs. It has been postulated that ROS stimulates antioxidant adaptation and that adaptive responses are depressed with age. However, aging is associated with increased tissue levels of antioxidants (reviewed by (Ji *et al.*, 1998). These include the antioxidant enzymes SOD, catalase and glutathione peroxidase; and glutathione metabolism regulators glutathione-S-transferase and glutathione reductase. Antioxidant adaptation is regulated at both the translational and posttranslational levels (Hollander, Bejma *et al.*, 2000). Therefore, although antioxidants are maintained or increased with age, altered posttranslational modifications may result in inefficient antioxidant activity.

In addition, HSP function may be reduced in aged individuals. In healthy young individuals, ROS influences the expression of HSP25, HSP72 and HSP60, as well as other members of the HSP family (Spiers *et al.*, 2000; Naito *et al.*, 2001; McArdle *et al.*, 2002). HSPs are induced following oxidative stresses such as strenuous exercise. This induction is severely blunted in aged individuals (Vasilaki *et al.*, 2002).

Consequently, some aspects of age-related muscle dysfunction may be mediated by a failure to induce HSPs. This concept is supported by the observation that long-lived *Caenorhabditis elegans* mutants show increased HSP expression and stress resistance (Walker *et al.,* 1998; Sampay *et al.,* 2000).

The second process affected by oxidative signaling is the regulation of muscle protein content. A rise in sarcoplasmic ROS has been shown to activate redox-sensitive kinases and nuclear factor-κB (NF-κB), one of several transcription factors that exhibit oxidative activation (Li *et al.,* 1998; Li *et al.,* 1999). In turn, persistent activation of NF-κB leads to progressive loss of muscle protein (Li and Reid, 2000). NF-κB activation is increased with age, however it has not been demonstrated that this over-activation leads to muscle loss in aged individuals (Zhang *et al.,* 2004).

7.6 Perturbing the System: Effects of Nutrition

Anorexia and malnutrition are common elements of the aging process. Nutritional interventions are therefore advocated to slow or reverse the progression of sarcopenia (Parise and Yarasheski, 2000; Hebuterne *et al.,* 2001). The primary goal has been to increase amino acid availability and thereby facilitate resynthesis of muscle protein. In addition, aged muscle may be especially susceptible to the loss of nutritional antioxidants (Ji, *et al.,* 1998). Therefore, an increased intake of antioxidant nutrients might partially counteract the ROS and NO contribution to age-related losses (Weindruch, 1995). This concept has not been confirmed. In principle, however, age-related oxidative damage could be blunted in muscle cells by increasing the availability of antioxidant nutrients such as reduced thiols, vitamins E and C, β-carotene, carotenoids and α-lipoic acid. Nutritional antioxidants may also enhance NO signaling (Carr and Frei, 2000), potentially restoring age-related losses. Antioxidant supplementation is not likely to affect function in the basal state since muscles of aged individuals are not overtly deficient in antioxidant nutrients. However, supplements may be important for aged individuals who are physically active or chronically ill. Exercise and inflammation increase the oxidant burden in muscle, and this increase may overwhelm normal antioxidant levels.

In addition to nutritional supplements for aged individuals, an altered pattern of food intake for younger individuals may delay aging. This altered pattern involves caloric restriction (CR). Studies show that CR delays aging in insects, fish, spiders, water fleas, rats and mice (McCay *et al.*, 1989; Weindruch, 1996; Lass *et al.*, 1998). In rodents, CR attenuates the loss of muscle mass that accompanies age-related disease such as diabetes and hypertension (Weindruch and Sohal, 1997; Weindruch and Walford, 1998). In addition, it has been shown that CR reduces age-related ROS production and oxidative damage (Sohal and Weindruch,1996; Leeuwenburgh *et al.*, 1997; Lass *et al.*, 1998; Zainal *et al.*, 2000; Gredilla *et al.*, 2001; Drew *et al.*, 2003).

7.7 Perturbing the System: Effects of Cytokines

Loss of muscle function with aging has long been linked to increased circulating cytokines (Cannon, 1995; Cannon, 1998; Mackinnon, 1998; Pahor and Kritchevsky, 1998; Kotler, 2000). Potential mediators include interleukin 1 (IL-1), interleukin 6 (IL-6) and tumor necrosis factor-α (TNF- α). These cytokines are elevated in the circulation of elderly individuals (Greiwe *et al.*, 2001). Basal levels are further elevated by inflammatory disease and are transiently increased following strenuous exercise (Reid and Li, 2001). The effects of TNF- α are best understood. TNF- α stimulates cellular ROS production and induces loss of muscle function. The signaling events that regulate this process involve binding of TNF- α to sarcolemmal receptors and stimulation of ROS production by muscle mitochondrial (Reid and Li, 2001). TNF- α-stimulated ROS activates two events that have additive effects on muscle performance. Within hours after TNF- α exposure, ROS disrupt contractile regulation and weaken the muscle. With longer exposure, TNF-α-induced ROS act via NF-κB and p38 MAPK to up-regulate the ubiquitin/proteasome pathway, accelerate protein breakdown and diminish muscle mass (Li *et al.*, 2005). Given their potential importance in age-related muscle loss, catabolic cytokines are major targets for therapeutic development. Interventions to inhibit cytokine effects on muscle are in early stages of clinical testing (Kotler, 2000). If proven safe and effective, such compounds could lead to treatments for muscle loss in the elderly.

7.8 Perturbing the System: Effects of Training

Exercise is among the most effective interventions for loss of function in aging muscle. Although the conditioning process is prolonged in aged-muscle, a well-designed strength-training program can safely increase muscle mass, strength and function (Roth *et al.*, 2000; Hebuterne *et al.*, 2001). By definition, the exercise bouts that are used for training will transiently elevate tissue ROS and NO levels. In young adults, such training evokes adaptive responses in skeletal muscle fibers. These include an increase in antioxidant enzyme activity that renders the muscle more resistant to oxidative stress (Powers *et al.*, 1999). HSP levels are also up-regulated in trained muscle (McArdle and Jackson, 2000; Naito *et al.*, 2001), which further contributes to cytoprotection. Concerns have been raised that aging may limit the capacity of muscle to adapt in this manner (Spiers *et al.*, 2000). Although HSP expression appears to be blunted with age, the increases in antioxidant enzyme activities that occur in muscles after training are generally similar between adults and aged individuals (Ji *et al.*, 1998). Consequently, with a moderate training regimen, the adaptive responses that are still intact may be able to compensate for those that are deficient. Recent studies with aged mice and rats show they are able to condition, but that the process is prolonged (Gosselin, 2000; Brooks *et al.*, 2001). In addition to antioxidant up-regulation, which blunts ROS activity, another important result of exercise training for the elderly is a decrease in circulating TNF-α levels (Greiwe *et al.*, 2001), thus reducing one of the contributors to ROS production in the elderly. These adaptive responses are expected to oppose the action of ROS, decreasing the activity of oxidant-activated pathways and improving muscle function.

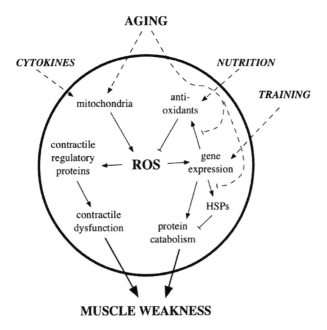

Fig. 7.4. Effects of aging on pathways regulated by ROS. *Circle* represents sarcolemma, pathways that lead to muscle weakness are shown. *Dashed lines* represent aging and other external factors, *solid lines* represent pathways leading to weakness. *Arrows* denote stimulation, *blunt ends* indicate inhibition.

7.9 Summary

The loss of muscle function that occurs with aging is likely to be mediated, at least in part, by muscle-derived ROS and NO (Fig. 7.4). It appears that ROS production during muscle contraction is exaggerated due to the loss of mitochondrial regulation that occurs with aging. It is also likely that proteins in aged skeletal muscle are more susceptible to oxidative modification, rendering muscle fibers more sensitive to oxidant effects. Additionally, repair and protein degradation mechanisms are less effective (Stadtman, 1990). Redox homeostasis of the aging myocyte is further disrupted by humoral factors, including increased exposure to catabolic cytokines. As our understanding of this biology matures, targeted interventions including nutrition, exercise and pharmacological agents are being developed that may lessen or reverse the deleterious effects of aging on skeletal muscle.

References

1. Alvarez, S. and Boveris, A. *Free Radic. Biol. Med.* **37**(2004), 1472-8.
2. Andrade, F. H. *et al., J. Physiol.* **509** (1998), 565-75.
3. Andrade, F. H. *et al., Faseb. J.* **15** (2001), 309-11.
4. Bejma, J. and Ji, L.L. *J. Appl. Physiol.* **87** (1999), 465-70.
5. Brooks, S.V. *et al., J. Gerontol. A. Biol. Sci. Med. Sci.* **56** (2001): B163-71.
6. Cannon, J.G. *J. Gerontol. A. Biol. Sci. Med. Sci.* **50** (1995), 120-3.
7. Cannon, J.G. *Ann. N Y Acad. Sci.* **854** (1998), 72-7.
8. Carr, A. and Frei, B. *Free Radic. Biol. Med.* **28** (2000), 1806-14.
9. Drew, B. *et al., Regul. Integr. Comp. Physiol.* **284** (2003), R474-80.
10. Dumont, P. *et al., FEBS Lett.* **502** (2001), 109-12.
11. Fielding, R.A. and Meydani, M. *Aging (Milano)* **9**(1997), 12-8.
12. Fulle, S. *et al., Neuromuscul. Disord.* **13** (2003), 479-84.
13. Fulle, S. *et al., Free Radic. Biol. Med.* **29** (2000), 1252-9.
14. Gianni, P. *et al., Exp. Gerontol.* **39** (2004), 1391-400.
15. Gosselin, L.E. *J. Appl. Physiol.* **88** (2000), 1254-8.
16. Gredilla, R. *et al., J. Bioenerg. Biomembr.* **33** (2001), 279-87.
17. Greiwe, J.S. *et al., Faseb. J.* **15** (2001), 475-82.
18. Harman, D. *J. Am. Geriatr. Soc.* **20** (1972), 145-147.
19. Hebuterne, X. *et al., Curr. Opin. Clin. Nutr. Metab. Care.* **4** (2001), 295-300.
20. Hollander, J. *et al., Mech. Ageing Dev.* **116** (2000), 33-45.
21. Ji, L.L. *et al., Ann. N Y Acad. Sci.* **854** (1998), 102-17.
22. Kanski, J. *et al., Free Radic. Biol. Med.* **35** (2003), 1229-39.
23. Kayo, T. *et al., Proc. Natl. Acad. Sci. U S A* **98**(2001), 5093-8.
24. Kotler, D.P. *Ann. Intern. Med.* **133** (2000), 622-34.
25. Lass, A., *et al., Free Radic. Biol. Med.* **25** (1998), 1089-97.
26. Lawler, J.M., *et al., Am. J. Physiol.* **272** (1997), E201-7.
27. Leeuwenburgh, C. *et al., Arch. Biochem. Biophys.* **346** (1997), 74-80.
28. Li, Y. *et al., Faseb. J.* (2005), **In Press**.
29. Li, Y.P. *et al., Antioxid. Redox. Signal.* **1** (1999), 97-104.
30. Li, Y.P. and Reid, M.B. *Am. J. Physiol. Regul. Integr. Comp. Physiol.* **279** (2000), R1165-70.
31. Li, Y.P. *et al., Faseb. J.* **12** (1998), 871-80.
32. Mackinnon, L.T. *Int. J. Sports Med.* **19** (1998), S205-9; discussion S209-11.

33. McArdle, A. and Jackson, M.J. *J. Anat.* **197** (2000), 539-41.
34. McArdle, A. *et al., Ageing. Res. Rev.* **1** (2002), 79-93.
35. McCay, C.M. *et al., Nutrition.* **5** (1989), 155-71; discussion 172.
36. Naito, H. *et al., Med. Sci. Sports Exerc.* **33** (2001), 729-34.
37. Navarro, A. *et al., Front. Biosci.* **6** (2001), D26-44.
38. Pahor, M. and Kritchevsky, S. *J. Nutr. Health Aging.* **2** (1998), 97-100.
39. Pansarasa, O. *et al., Free Radic. Biol. Med.* **27** (1999), 617-22.
40. Pansarasa, O. *et al., Free Radic. Res.* **33** (2000), 287-93.
41. Parise, G. and Yarasheski, K.E. *Curr. Opin. Clin. Nutr. Meta. Care.* **3** (2000), 489-95.
42. Patwell, D.M. *et al., Free Radic. Biol. Med.* **37** (2004), 1064-72.
43. Pesce, V. *et al., Free Radic. Biol. Med.* **30** (2001), 1223-33.
44. Powers, S.K. and Hamilton, K. *Clin. Sports Med.* **18** (1999), 525-36.
45. Powers, S.K. *et al., Med. Sci. Sports Exerc.* **31** (1999), 987-97.
46. Reid, M.B. *Med. Sci. Sports Exerc.* **33** (2001), 371-6.
47. Reid, M.B. *J. Appl. Physiol.* **90** (2001), 724-731.
48. Reid, M.B. *et al., J. Appl. Physiol.* **75** (1993), 1081-7.
49. Reid, M.B. and Li, Y.P. *Acta. Physiol. Scand.* **171** (2001), 225-232.
50. Renault, V. *et al., Exp. Gerontol.* **37** (2002), 1229-36.
51. Renault, V. *et al., Cell.* **1** (2002), 132-9.
52. Richmonds, C.R. *et al., Mech. Ageing Dev.* **109** (1999), 177-89.
53. Richmonds, C.R. and Kaminski, H.J. *Mech. Ageing Dev.* **113** (2000), 183-91.
54. Roth, S.M. *et al., J. Nutr. Health Aging.* **4** (2000), 143-55.
55. Sampayo, J.N. *et al., Ann. N Y Acad. Sci.* **908** (2000), 324-6.
56. Schoneich, C. *et al., Mech. Ageing Dev.* **107** (1999), 221-31.
57. Sen, C.K. and Packer, L. *Faseb. J.* **10** (1996), 709-20.
58. Sohal, R.S. and Weindruch, R. *Science* **273** (1996), 59-63.
59. Spiers, S. *et al., Ann. N Y Acad. Sci.* **908** (2000), 341-3.
60. Stadtman, E.R. *Biochemistry* **29** (1990), 6323-31.
61. Vasilaki, A. *et al., Muscle Nerve* **25** (2002), 902-5.
62. Viner, R.I. *et al., Biochem. J.* **340** (1999), 657-69.
63. von Zglinicki, T. *et al., Exp. Cell Res.* **220** (1995), 186-93.
64. Walker, G.A. *et al., Ann. N Y Aca. Sci.* **851** (1998), 444-9.
65. Wallace, D.C. *Novartis. Found. Symp.* **235** (2001), 247-63; discussion 263-6.
66. Weindruch, R. *J. Gerontol. A. Biol. Sci. Med. Sci.* **50** (1995), 157-61.
67. Weindruch, R. *Sci. Am.* **274** (1996), 46-52.
68. Weindruch, R. and Sohal, R.S. *N. Engl. J. Med.* **337** (1997), 986-994.
69. Weindruch, R. and Walford, R.L. *The Retardation of Aging and Disease by Dietary Restriction.* (Springfield, IL, CC Thomas, 1998).
70. Zainal, T.A. *et al., Faseb. J.* **14** (2000), 1825-36.
71. Zhang, J. *et al., Exp. Gerontol.* **39** (2004), 239-47.

CHAPTER 8

AGING, EXERCISE, ANTIOXIDANTS, AND CARDIOPROTECTION

John Quindry[1] and Scott Powers[2]

[1]*Appalachian State University, Boone, NC;* [2]*University of Florida, Gainsville, FL*

8.1 Introduction

Heart disease is the leading causes of death in the United States. Among the various forms, ischemic heart disease, or ischemia-reperfusion (IR) injury, accounts for the majority of all heart disease deaths (AHA, 2004). The magnitude of myocardial IR-induced myocardial injuries can vary from minor to severe damage, with the latter often resulting in permanent disability and death. Although IR injury can occur in persons of any age, the likelihood of suffering an IR event increases as a function of age (AHA, 2004). Given the worldwide prevalence and financial burden of IR injury, discovering practical and cost effective countermeasures against IR-induced heart disease is of critical importance.

In this regard, physical exercise has proven to be the most effective means of protecting the myocardium against IR-induced cardiac injury. Indeed, it now is well established that regular endurance exercise (e.g., treadmill exercise) provides cardioprotection against IR injury in both young and old animal (Demirel *et al.,* 1998, 2001; Hamilton *et al.,* 2001; 2003; 2004; Lennon *et al.,* 2004a-c, Powers *et al.,* 1998; Quindry *et al.,* 2005, Starnes *et al.,* 2003; Taylor *et al.,* 1999). This chapter will summarize the current knowledge about the effects of age on myocardial IR injury. Further, evidence will be presented detailing that endurance exercise is an effective countermeasure against IR injury throughout the lifespan. Finally, we will also discuss the potential mechanisms responsible for exercise-induced cardioprotection.

8.2 Overview of Myocardial IR Injury

Myocardial injury caused by IR is a serious clinical problem. Interruption of blood flow to the contracting cardiac muscle, termed ischemia, rapidly results in cellular disruption due to a severe energy supply/demand mismatch (Kloner and Jennings, 2001). Paradoxically, restoration of blood flow to the myocardium, called reperfusion, results in additional injury to cardiac myocytes (Vetterlien *et al.*, 2003). The specific mechanisms that contribute to IR injury are complex and involve numerous mediators that act in concert. Nonetheless, the essential factors leading to IR-induced cellular dysfunction and death have been delineated and include production of free radicals, calcium overload, protease activation (i.e. calpain), deleterious membrane alterations, and inflammation (Bolli and Marban, 1999). For both ischemia and reperfusion, the magnitude of IR injury accrues in a time-dependent fashion (Downey, 1990). As a generalization, IR damage is described in three distinct levels of injury based on the duration of ischemia (Fig. 8.1). Following 1 to 5 minutes of ischemia the first quantifiable level of injury is arrhythmia generation. Occurring primarily during both ischemia and reperfusion, ventricular tachycardia and fibrillation can occur in the absence of permanent decrements in contractile function. The second level of IR-induced myocardial injury, known as stunning, occurs during reperfusion after an ischemic period of 5 to 20 minutes. Myocardial stunning is characterized by ventricular contractile deficits that occur without cell death. Recovery of contractility in the stunned myocardium typically occurs over the course of several days following the ischemic event (Bolli and Marban, 1999). The third and highest level of IR injury occurs when the ischemic duration extends beyond 20 minutes. Under these circumstances, IR promotes myocardial cell death (i.e. myocardial infarction). The magnitude of the infarction and left ventricle pump dysfunction is generally dependent upon the mass of myocardium affected and the duration of the ischemic event (Downey, 1990). Clinical implications of infarction are broad as myocardial infarction is the leading cause of cardiovascular morbidity and mortality (AHA, 2004).

Levels of myocardial injury

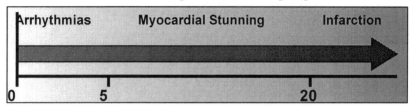

Fig. 8.1. A schematic illustration of the IR-induced myocardial injury continuum. Myocardial injury due to ischemia accrues in a time-dependent manner with separation between the three injury levels occurring at the listed time approximations.

8.3 Myocardial IR: Necrotic and Apoptotic Cell Death

Recent evidence reveals that IR-induced myocardial cell death is not limited to necrosis but also includes apoptotic cell death (Kloner and Jennings, 2001; McCully *et al.*, 2004). Apoptosis can be generally defined as programmed cell suicide utilizing specific mechanisms of initiation and execution in an energy dependent manner. Morphologically, apoptosis is characterized by nuclear chromatin condensation, loss of nuclear and cellular volume and finally blebbing of the nuclear and cellular membranes (Kloner and Jennings, 2001). Thus, apoptotic cell death is pathologically different from necrosis in that the former is a self-destructive response to unfavorable cellular conditions while the latter involves cell death via an external stimulus and cell rupture (Kloner and Jennings, 2001). There are numerous internal (i.e. noxious cellular environment) and external (i.e. cell death receptor stimulation) stimuli responsible for pathologic apoptosis. Activation of apoptotic death receptors on the sarcolemma includes activation multiple signaling pathways including NFKB, JNK and FADD. Alternately, internal mediators of apoptosis are inevitably mitochondrion-dependent resulting in an apoptotic signaling cascade through release of either cytochrome c or apoptosis inducing factor (AIF) from the mitochondrion (Haunstetter and Izumo, 2001). Importantly, cell death receptor and mitochondrial inducers of myocardial apoptosis result in activation of cysteine-aspartate dependent proteases, called caspases. Downstream caspase activation leading to apoptosis ultimately necessitates activation of caspase-3 as the final mechanism of apoptosis induction. Once activated, caspase-3 cleaves the DNA repair enzyme poly (ADP)-ribosylating protein (PARP) resulting in over activation. Once PARP is over activated, cellular energy stores are further depleted and apoptosis is initiated.

With respect to internal stimuli for myocardial apoptosis, growing evidence indicates that IR-induced necrotic and apoptotic forms of cell death occur in tandem and involve similar stimuli such as oxidative stress and Ca^{2+} overload (Fig. 8.2) (Kumar and Jugdutt, 2003). Since cardiac myocytes do not contain satellite cells and are incapable of cell division, prevention of IR-induced cell death is critical to the preservation of cardiac function following IR events. The next two sections will overview various aspects of free radical and calcium-mediated processes as mechanisms related to IR-induced myocardial injury and death.

8.4 Oxidative Stress and Myocardial IR

Free radicals are chemically reactive molecules or molecule fragments with an unpaired electron in the outer orbital. The term oxidative stress is often used to generically describe free radical-mediated oxidative damage to molecules (i.e. proteins or lipids). One of the most physiologically relevant oxygen-derived free radicals, superoxide, is produced by a univalent reduction of molecular oxygen. Within cellular environments, superoxide is relatively benign compared to other radicals. Nonetheless, through spontaneous and enzyme-catalyzed reactions, superoxide is easily converted to other reactive oxygen species (ROS) including the hydroxyl radical, hydrogen peroxide, peroxyradicals, and peroxynitrite. These derivative ROS are extremely cytotoxic by promoting free radical propagation reactions (Halliwell *et al.*, 1993; Yu, 1994).

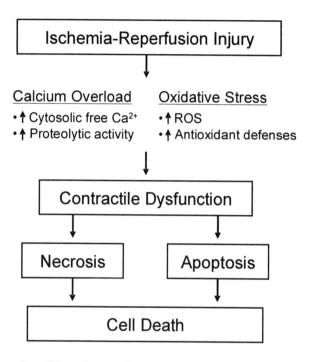

Fig. 8.2. An overview of the major contributors to ischemia-reperfusion injury resulting in contractile dysfunction, and potentially, cell death via necrotic and apoptotic processes. The observation that hearts from exercised animals are protected against IR-induced contractile dysfunction and cell death is primarily attributed to an improved capacity to control oxidative stress and calcium overload.

Free radicals and ROS can produce several types of cellular damage including oxidative modification to lipids, protein, and nucleic acids. Oxidative damage to lipids (lipid peroxidation) occurs when radical species react with polyunsaturated fatty acids in membranes. Lipid peroxidation results in propagation reactions cumulating in the formation of new radicals (i.e. peroxyradicals, peroxynitrite) (Halliwell *et al.*, 1993; Yu, 1994). Lipid peroxidation of membrane fatty acids leads to altered membrane fluidity, increased permeability, and altered membrane function due to damage to membrane components. Damage to cellular proteins by free radicals occurs primarily in amino acids with exposed thiol groups. The resulting damage to ROS-sensitive amino acids involves carbonyl formation on their side chains and a loss of protein function. The implications of diffuse cellular protein oxidative damage are significant as widespread oxidative stress will lead to damage of enzymes, receptors and proteins responsible for transport, structural integrity, and contraction in cardiac muscle (Bolli, 1988; 1999; Downey, 1990; Halliwell *et al.*, 1993). Finally, oxidative damage to DNA can produce mutations to the genetic code both directly and indirectly during attempts at DNA repair and replication. Hence, oxidative damage to DNA can lead to cellular dysfunction via the expression of dysfunctional proteins. Moreover, DNA strand breakage can result in activation of PARP and subsequent apoptotic cell death as described previously (Halliwell *et al.*, 1993; Lucchesi, 2001).

Ample evidence exists to indicate that free radical stress occurs during both ischemia and reperfusion, though the majority of IR-induced oxidative stress arguably occurs during reperfusion. Primary sources of free radicals in contracting cardiac muscle are the respiratory electron carriers within the mitochondria, particularly complex III. During ischemia, mitochondrial respiration (electron transfer) becomes arrested and thus oxygen radical formation may be reduced slightly. Upon reperfusion, however, a significant free radical burst can occur as respiratory electron carriers are re-oxidized (Lucchesi, 2001). This scenario gains physiological significance during periods of intermittent IR often incurred by individuals suffering from cardiovascular disease. Further, cytosolic calcium overload during IR results in activation of xanthine oxidase and NADPH/NADH oxidase, both of which enzymatically produce superoxide (Halliwell *et al.*, 1993). Upon reperfusion, extravasisation of neutrophils to the ischemic region is a critical component of the inflammation response to IR that also leads to significant oxidative stress. Further, as part of the inflammation response of IR, superoxide production is also increased by activation of arachidonic acid and cyclooxygenase-2 (COX-2) while the inducible isoform of nitric oxide synthase (iNOS) activation leads to nitric oxide (NO) production (Bolli, 1999; Halliwell *et al.*, 1993; Lucchesi, 2001). In aggregate, the oxidative stress imposed during IR can overwhelm the endogenous antioxidant defenses leading to cellular dysfunction and death.

8.5 Calcium Overload and Myocardial IR

Disturbances in cellular calcium homeostasis are arguably responsible for much of the IR-induced myocardial injury. Direct measurement of myocardial calcium concentration indicates that calcium overload occurs during reperfusion with ischemic periods as brief as 9 to 15 minutes (Carrozza *et al.*, 1992; Dhalla *et al.*, 2001). As a brief overview of the processes leading to calcium overload during IR, the cessation of oxygen delivery during ischemia quickly leads to cytosolic hydrogen ion accumulation and a decreased intracellular pH. As cellular metabolism shifts from oxidative to glycolytic pathways, the decrease in pH is exacerbated. Sarcolemmal sodium-hydrogen exchanger activity is increased dramatically to meet the increased pH challenge, ultimately leading to sodium accumulation within the cytosol. Secondary to ATP depletion, the intracellular sodium concentration is further increased as Na+-K+ ATPase activity is slowed or terminated. In an effort to preserve the rapidly diminishing membrane potential, Na^+-Ca^{2+} antiporter activity is increased resulting in calcium overload. Elevated cytosolic free calcium impairs cardiac contractility by several mechanisms including calcium-activated proteases (i.e. calpain), impaired excitation-contraction coupling, and decreased responsiveness of the contractile elements to calcium (Bolli, 1999; Dhalla *et al.*, 2001). Further cellular disruption due to calcium overload includes activation of phospholipases, enzymatic free radical production, disruption of mitochondrial energy production, and a loss of cellular integrity.

While the mechanisms of calcium overload and oxidative stress explain many of the pathological events leading to myocardial IR-injury, these processes are likely interrelated (Bolli, 1999). For instance, excessive free radical production during IR can alter membrane fluidity and thereby alter permeability. This can impede Na+ extrusion through several ion transport systems thereby exacerbating the cytosolic Na+ accumulation and subsequent calcium influx. Likewise, calcium overload also results in increased production of ROS via several different pathways.

8.6 Senescence and Increased Susceptibility of the Myocardium to IR Injury

A growing volume of evidence indicates that numerous cellular and functional changes occur over the course of a life span that renders the heart susceptible to IR injury (AHA, 2004). Importantly, many of these alterations are associated with alterations in gene expression rather than processes of degeneration (Lakatta and Sollot, 2002). Ongoing research is currently directed at delineating the mechanisms underlying the increased susceptibility to IR injury with age and will be discussed in the subsequent sections.

The mechanisms of IR-induced myocardial dysfunction discussed above are exacerbated with senescence. First, mitochondria from aged hearts produce more ROS than mitochondria from younger hearts resulting in diffuse mitochondrial oxidative damage (Lesnefsky *et al.,* 1994). Thus, increased mitochondrial ROS production by aged hearts ultimately lead to compromised energy generation capacity, further endangering the cardiac myocyte in the event of IR. Second, cardiac tissue in aged animals has a diminished ability to control both calcium release and reuptake. As compared to hearts from younger animals, aged hearts experience slower inactivation of L-type Ca^{2+} channels. Further, senescence is associated with diminished levels of sarcoplasmic/endoplasmic reticulum calcium ATPase (SERCA2a) mRNA and receptor density. Collectively, these age-related changes in calcium regulation increase excitation-contraction calcium transients, and thus, an increased likelihood for IR-induced calcium overload (Lakatta and Sollott, 2002).

In addition to augmentation of mediators of IR damage, the senescent heart also appears to be more susceptible to oxidative injury than hearts from younger counterparts. For instance, unfavorable alterations in glutathione redox status with age may render the myocardium more susceptible to free radical injury (Rebrin *et al.*, 2003).

Similarly, normal aging is associated with an increased ratio of mitochondrial and sarcolemmal ω-6 PUFA/ω-3 PUFA ratios and decreased cardiolipin content. Age associated alterations in the ω-6 PUFA/ω-3 PUFA and cardiolipin membrane content results in an increased susceptibility to free radical and Ca^{2+} overload damage in the event of an ischemic stress. Further, cellular damage subsequently results through a loss in membrane potential and initiation of caspases and other apoptotic signaling mechanisms (Pepe, 2000).

Previous reports suggest that, compared to young rats, hearts from senescent animals are more susceptible to IR-induced myocardial injury and necrotic cell death (Headrick, 1998; Lesnefsky *et al.*, 1994; Liu *et al.*, 2002; Tani *et al.*, 1997). Recent studies reveal that IR-induced myocardial cell death is not limited to necrosis but also includes apoptotic cell death (Chen *et al.*, 2002; Liu *et al.*, 2002). Indeed, growing evidence indicates that IR-induced necrotic and apoptotic cell death occurs in tandem, and that similar stimuli (e.g., oxidative stress and Ca^{2+} overload) may contribute to both forms of cell death (Kumar and Jugdutt, 2003). However, to date, only two reports have investigated the influence of age on IR-induced myocardial apoptosis (Liu *et al.*, 2002; Quindry, 2005). Findings from both studies indicate that hearts from aged rats are more susceptible to I-R-induced apoptosis than their young counterparts. In contrast to the effects of aging, moderate intensity endurance exercise has a beneficial effect on gene expression that results in cardioprotection and is the focus of the rest of this chapter.

8.7 Exercise-Induced Protection against Myocardial IR Injury

It is well established that aerobic exercise training results in improved myocardial tolerance to IR challenges (Demirel *et al.*, 2001; Powers *et al.*, 1998; Taylor *et al.*, 1999). Exercise-mediated cardioprotection is afforded to both moderate-duration ischemia (i.e. 5–20 minutes of ischemia resulting in myocardial stunning) and long-duration ischemia (i.e. 20–60 minutes of ischemia resulting in myocardial infarction) (Demirel *et al.*, 2001; Powers *et al.*, 1998; Taylor *et al.*, 1999). Studies from our laboratory and others indicate that both long-term exercise training (i.e. 10 weeks) (Demirel *et al.*, 1998; Harris and Starnes, 2001) and short-term endurance exercise exposure (i.e. 1–10 days) (Demirel *et al.*, 2001, Hamilton *et al.*, 2001; 2004; Lennon *et al.*, 2004a-c, Yamashita *et al.*, 1999) provide similar cardiac protection against IR insults. Further, we have recently identified that 8 days of either moderate-intensity (i.e. ~50% VO$_2$max) and high-intensity (i.e. ~75% VO$_2$max) exercise provided similar levels of cardioprotection against myocardial stunning of isolated perfused rat hearts (Fig. 8.3) (Lennon *et al.*, 2004a). In a separate study, employing an identical IR-stunning insult, our group demonstrated that the time course of exercise-induced cardioprotection afforded by 3 days of treadmill exercise persists for between 9 and 18 days following exercise cessation (Fig. 8.3) (Lennon *et al.*, 2004b). Finally, growing evidence indicates that exercise-induced cardioprotection is an age-independent phenomenon (Quindry *et al.*, 2005; Starnes *et al.*, 2003).

8.8 Mechanisms for Exercise-Induced Cardioprotection

Presently, the mechanisms responsible for exercise-induced myocardial protection against IR injury are not fully known. However, based on knowledge of the key components of IR injury, several theoretical mechanisms could explain the cardioprotective effect of endurance exercise. These include anatomical alterations in the coronary arteries (i.e. collateral coronary circulation), induction of myocardial heat shock proteins, improved myocardial antioxidant capacity, maintenance cellular calcium homeostasis, and possibly, increases in other unknown cytoprotective proteins.

8.9 Exercise and Collateral Coronary Circulation

The development of collateral coronary circulation may occur in some animal species in response to prolonged (i.e. months to years) endurance exercise training. However, the cardioprotective benefits of short-term exercise are not attributed to alterations in collateral circulation (Yamashita *et al.*, 1999). Thus, by a process of elimination, it seems likely that the cardioprotection observed in response to short-term exercise is due to myocardial expression of cardioprotective molecules. Indeed, this is a more biologically logical explanation for exercise-induced protection considering the previously discussed time-course for cardioprotective acquisition with exercise and the disappearance of cardioprotection upon cessation (Lennon *et al.*, 2004b).

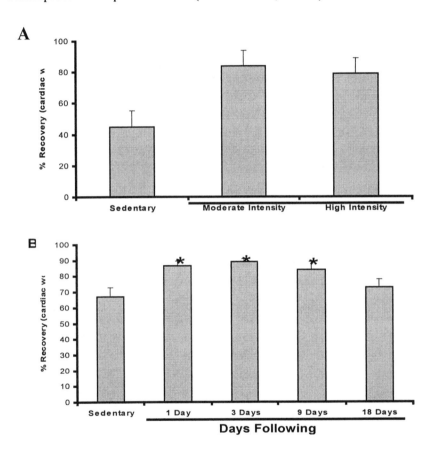

Fig. 8.3. Panel A – both moderate (~50% VO₂max) and high-intensity (~75% VO₂max) treadmill exercise provide similar levels of cardioprotection against a myocardial stunning insult in isolated perfused hearts. Panel B – 3 days of exercise induce cardioprotection against contractile deficits as a result of IR stunning. Cardioprotection persists for greater than 9 days following the cessation of exercise.

8.10 Exercise, Heat Shock Proteins, and Myocardial IR Injury

Cellular proteins play a critical role in maintaining cellular homeostasis and IR-induced damage to constitutive proteins and/or impaired protein synthesis result in a loss of cellular homeostasis. Thus, in response to a variety of stressors, the cell responds by synthesizing a group of proteins known as "heat shock proteins" (HSP). Stimuli responsible for HSP up-regulation include alterations in cellular redox status, increases in cell temperature, and acute damage to constitutive proteins (e.g. contractile elements within cardiac tissue); conditions also common to acute endurance exercise (Demirel *et al.*, 2003; Gray *et al.*, 1999; Hamilton *et al.*, 2001; Hutter *et al.*, 1994). While several HSPs (e.g. HSP-10, HSP-27, HSP-40, HSP-60, and HSP-90) provide cellular protection against stressors such as IR, evidence indicates that HSP-72 can be protective against IR-induced injury (Gray *et al.*, 1999).

Several investigative groups have examined the specific contribution of exercise-induced increases in myocardial HSPs to cardioprotection against IR injury. To identify the effect of exercise-induced elevations in core temperature on myocardial HSP content, Hamilton *et al.* (2001) exposed rats to treadmill running to both hyperthermic (25°C ambient temperature resulting in an exercise-induced increase in core temperature to ~41°C) and euthermic conditions (4°C ambient temperature resulting in no rise in core temperature during exercise). The results revealed that exercise in the cold prevented an increase in myocardial HSP72 whereas exercise in a warm room resulted in significant increases in myocardial levels of HSP-40, HSP-72, and HSP-90 (Hamilton *et al.*, 2001). Interestingly, however, animals exercised in both the cold and warm conditions (independent of myocardial HSP levels) demonstrated similar levels of cardioprotection against IR-induced myocardial stunning (Hamilton *et al.*, 2001). These findings reveal that HSPs are not a critical component of exercise-induced cardioprotection against myocardial stunning and support previous work by Taylor *et al.* (1999) indicating that HSP-72 is not an essential to exercise-induced cardioprotection against short-duration ischemia.

8.11 Exercise, Antioxidant Capacity, and Myocardial IR Injury

The myocardium contains numerous endogenous mechanisms of protection against ROS including enzymatic and non-enzymatic antioxidants. Primary enzymatic antioxidants include superoxide disumtase (SOD), catalase (CAT), and glutathione peroxidase (GPx). In mammals, SOD exists in cytosolic and mitochondrial isoforms. The cytosolic isoform requires CuZn as a cofactor whereas Mn is the cofactor in the mitochondrial isoform. Important non-enzymatic antioxidants in the cell include reduced glutathione (GSH), vitamin E, and vitamin C (Powers *et al.*, 1999).

In general, enzymatic and non-enzymatic antioxidants act in concert to convert ROS to less reactive forms. SOD, CAT, and GPx defend against ROS by quenching superoxide and hydrogen peroxide. Specifically, SOD converts superoxide to hydrogen peroxide while CAT and GPx convert hydrogen peroxide to oxygen and water. GPx utilizes GSH as a reducing equivalent producing oxidized glutathione (GSSG). In the presence of aqueous and lipid phase antioxidants, such as vitamins C and E, glutathione reductase can re-reduce GSSG to the non-oxidized form. Conversely, GSH can assist with vitamin C and E recycling in addition to direct ROS removal (Powers *et al.*, 1999).

The effect of exercise on primary antioxidant enzyme activities within the myocardium has been widely investigated. Most investigations report that myocardial GPx activity is not affected by exercise (Criswell *et al.*, 1993; Demirel *et al.*, 2001; Powers *et al.*, 1993). With respect to the effect of exercise on myocardial CAT, results are equivocal. Some studies demonstrate a rise in myocardial CAT activity (Lennon *et al.*, 2004a-c; Starnes *et al.*, 2003) whereas others report no increase (Demirel *et al.*, 2001; Hamilton *et al.*, 2001; 2003; 2004). Alternately, it is widely held that exposure to endurance exercise results in a rapid increase in myocardial levels of MnSOD. Our laboratory has demonstrated repeatedly that as few as 3 days of moderate intensity treadmill elevates myocardial MnSOD activity (Demirel *et al.*, 2001; Hamilton *et al.*, 2001; 2003; 2004; Lennon *et al.*, 2004a-c). In the case of GSH, however, it has been demonstrated that chronic exercise training (e.g., weeks to months) is required to elevate myocardial levels of the antioxidant (Husain and Hazelrigg, 2002; Powers *et al.*, 1998). At this time, little is known about the effects of exercise on other potentially important redox systems in the heart (i.e., thioredoxin, peroxiredoxin reductase, hemeoxygenase, etc.).

Based on the critical role that ROS play in IR-injury, several investigators have hypothesized that exercise-induced increases in cardiac antioxidants contribute to exercise-induced cardioprotection (Demirel *et al.*, 2001; Hamilton *et al.*, 2003; 2004; Lennon *et al.*, 2004c; Powers *et al.*, 1998; Yamashita *et al.*, 1999). To date, the strongest evidence to directly link exercise-induced increases in myocardial antioxidants and exercise-induced cardioprotection involves MnSOD. In this regard, several studies have employed the use of antisense oligonucleotide (ASON) technology to specifically block the exercise-induced increase in myocardial MnSOD (Hamilton *et al.*, 2005; Hamilton *et al.*, 2004; Lennon *et al.*, 2004c; Yamashita *et al.*, 1999). The first study, performed by Yamashita, *et al.* (1999), examined the role that exercise-induced increases in myocardial MnSOD plays in exercise-induced cardioprotection against IR-induced infarction. The results revealed that prevention of exercise-mediated increases in myocardial MnSOD resulted in the abolition of cardioprotection against IR-induced infarction. More recently, a study from our laboratory confirmed these findings and demonstrated that exercise-induced elevations in MnSOD plays an important protective role against IR-induced cardiac arrhythmias (Hamilton *et al.*, 2004). Moreover, a recent investigation from our laboratory demonstrated that exercise-induced increases in myocardial MnSOD activity is, at least partially responsible for the infarct-sparing effect of exercise against both necrotic and apoptotic cell death (Hamilton *et al.*, 2005). In contrast to these investigations, Lennon *et al.* (2004c) demonstrated that MnSOD was not essential for the exercise-induced cardioprotection against *in vitro* myocardial stunning-induced contractile deficits. Exercise-induced prevention of IR-mediated apoptosis will be discussed further in a subsequent section.

8.12 Exercise, Calcium Overload, and Myocardial IR Injury

The contribution of calcium overload to IR-injury is undeniable. Unfortunately, there is a paucity of evidence demonstrating that exercise-induced cardioprotection includes exercise-mediated alterations in myocardial calcium handling during IR. Bowles *et al.*, (1994) demonstrated that myocardial stunning doubled cardiac Ca^{2+} uptake during reperfusion, while 8 weeks of exercise training completely attenuated the increased Ca^{2+} uptake during IR. A more recent study by our lab demonstrated in sedentary hearts exposed to IR stunning, that pharmacological inhibition of the calcium-activated protease, calpain, preserved cardiac function to levels observed in exercised hearts. In support, calpain activity was similar in the cardioprotected hearts from exercise and calpain-inhibited groups (French *et al.*, 2005). Whether exercise altered calpain activity via improved calcium handling and/or direct allosteric regulation of calpain is currently unknown. Nonetheless, French *et al.* (2005) demonstrated that in contrast to sedentary controls, sarcoplasmic/endoplasmic calcium ATPase (SERCA-2A) remained intact in exercise-trained hearts exposed to IR. This finding is consistent with the notion that compared to hearts from sedentary animals, exercise-trained hearts better maintain Ca^{2+} homeostasis during IR (French *et al.*, 2005). While these results are promising, further study is necessary to fully delineate the contribution of exercise-mediated alterations in calcium handling as a component of exercise-induced cardioprotection against IR injury.

8.13 Exercise, Age, and Cardioprotection against IR-Induced Apoptosis

Recent data reveal that, compared to young hearts, the senescent heart is more susceptible to IR-induced injury (Headrick, 1998; Lakatta and Sollott, 2002; Lesnefsky *et al.*, 1994; Pepe, 2000; Tani *et al.*, 1997). This observation is clinically relevant as the majority of myocardial IR events occur in individuals over the age of 65 years (AHA, 2004). Further, IR-induced myocardial cell death is not limited to necrosis but also includes apoptosis (Chen *et al.*, 2002; Liu *et al.*, 2002; Scarabelli *et al.*, 2001).

Indeed, IR-induced necrotic and apoptotic cell death are now known to occur in tandem, and that similar stimuli (i.e., oxidative stress and Ca^{2+} overload) may contribute to both forms of cell death (Kumar and Jugdutt, 2003). To date, only one study has been conducted to investigate exercise, age, and prevention of IR-induced apoptosis. A recent study by our laboratory investigated the contribution of exercise to prevent IR-induced apoptosis as a function of age. Our findings indicated that exercise did provide cardioprotection against IR-induced apoptosis. Importantly, the exercise-induced protection against apoptosis following IR-infarction was independent of age (Quindry *et al.*, 2005). These findings are in agreement with a previous study demonstrating that exercise provides age-independent cardioprotection against IR-stunning (Starnes *et al.*, 2003). Moreover, the findings of Quindry *et al.* (2005) corroborate a separate study performed by our lab that also demonstrated exercise-induced cardioprotection against IR-induced apoptosis (Hamilton *et al.*, 2005). The exact mechanisms for this exercise-induced protection against apoptosis remain unknown.

8.14 Exercise-Induced Cardioprotection: Future Directions

The need to identify specific mediators for exercise-induced cardioprotection is clearly present. The possibility exists that exercise promotes the expression of yet-to-be-identified cardioprotective mediators. Importantly, recent advances in biotechnology significantly improve the odds of identifying these previously unknown molecules. For instance, techniques such as DNA microarray analysis now permit the simultaneous measurement of thousands of gene transcripts in a single assay. With respect to exercise-induced alterations to the myocardium, our laboratory has performed a series of microarray experiments to investigate the effect of a single bout of endurance exercise (60 minutes of treadmill exercise ~60% VO_2max) on cardiac gene expression in young and aged rats. Results indicated that compared to hearts from sedentary animals, acute endurance exercise elicited a differential gene response. Further, the genetic response to exercise differed between epicardial and endocardial tissue in both young and old animals. Finally, a significant proportion of the altered gene transcripts in exercised hearts were expressed sequence tags (EST), or unidentified gene transcripts (Powers *et al.*, 2004). That this response was particularly notable in aged animals as compared to young and further emphasizes the potential for discovery of novel mediators of exercise-induced cardioprotection.

8.15 Summary

Table 8.1. Exercise is cardioprotective against ischemia-reperfusion injury independent of age. Among the cardioprotective attributes of exercise are improved redox status and control of calcium-related events during an ischemia-reperfusion insult.

Exercised Young	Exercised Old	Sedentary Young	Sedentary Old	Factors that affect outcome to myocardial ischemia-reperfusion injury
↓	↓	↑	↑	Oxidative Stress
↑	↑	↓	↓	Antioxidant Defenses
↓?	↓?	↑	↑	Calcium Overload
↓	↓	↑	↑	Calpain Activity

Exercise training decreases myocardial IR injury and improves cardiac contractile function following an IR-insult independent of age (Table 8.1). Recent evidence indicates that exercise protects against IR-induced cardiac arrhythmias, stunning, and infarction. Further, it is now clear that IR-induced necrotic and apoptotic forms of cell death are both attenuated in exercised hearts. Just as IR- injury is a multifaceted phenomenon, the cardioprotection afforded by exercise appears to be polygenic. Mediators of exercise-induced cardioprotection may include exercise-induced expression of HSPs, antioxidant enzymes, and potentially, control of calcium homeostasis and calcium mediated protease activity. Recent evidence, however, indicates that exercise-induced cardioprotection against myocardial stunning can be achieved independent of an increase in HSP-72. In addition, new findings reveal that MnSOD is at least partially responsible for the exercise-induced protection against IR-induced arrhythmia, cell death, and apoptosis, but not myocardial stunning. Finally, there is a strong possibility that unidentified protective molecules may also contribute to the cardioprotective response to exercise. Improving our basic understanding of the mechanisms responsible for exercise-induced protection against myocardial IR-injury will have broad implications for the prevention and treatment of patients predisposed to IR-insults.

References

1. American Heart Association, *Heart disease and stroke statistics - 2004 update.* American Heart Association: Dallas. (2004), 1-52.
2. Bolli, R., *J. Am. Coll. Cardiol.* **12** (1988), 239-49.
3. Bolli, R. and Marban, E. *Physiol. Rev.* **79** (1999), 609-34.
4. Bowles, D.K. and Starnes, J.W. *J. Appl. Physiol.* **76** (1994), 1608-14.
5. Carrozza, J.P., Jr. *et al., Circ. Res.* **71** (1992), 1334-40.
6. Chen, M. *et al., J. Biol. Chem.* **277** (2002), 29181-6.
7. Criswell, D. *et al., Med. Sci. Sports Exerc.* **25** (1993),1135-40.
8. Demirel, H.A. *et al., Med. Sci. Sports Exerc.* **30** (1998), 1211-6.
9. Demirel, H.A. *et al., J. Appl. Physiol.* **91** (2001), 2205-12.
10. Demirel, H.A. *et al., Am. J. Physiol. Heart Circ. Physiol.* **285** (2003), H1609-15.
11. Dhalla, N. *et al., Heart Physiology and Pathophysiology*, ed. N. Sperelakis. (Academic Press: San Diego, 2001), 949-965.
12. Downey, J.M. *Annu. Re. Physiol.* **52** (1990), 487-504.
13. French, J. *et al.,* (2005) *In Review.*
14. Gray, C.C. *et al., Int. J. Bioche. Cell Biol.* **31** (1999), 559-73.
15. Halliwell, B. and Gutteridge, J. *Free radicals in biology and medicine. Third ed.* (New York: Oxford University Press, 1999), 936.
16. Hamilton, K. *et al.,* (2005) *In Review.*
17. Hamilton, K.L. *et al., Am. J. Physiol. Heart Circ. Physiol.* **281** (2001), H1346-52.
18. Hamilton, K.L. *et al., Free Radic. Biol .Med.* **34** (2003), 800-9.
19. Hamilton, K.L. *et al., Free Radic. Biol. Med.* **37** (2004), 1360-8.
20. Harris, M.B. and Starnes, J.W. *Am. J. Physiol. Heart Circ. Physiol.* **280** (201), H2271-80.
21. Haunstetter, A. and Izumo, S. *Heart Physiology and Pathophysiology*, ed. Sperelakis. (Academic Press: San Diego, 2001), 927-947.
22. Headrick, J.P. *J. Mol.Cell. Cardiol.* **30** (1998), 1415-30.
23. Husain, K. and Hazelrigg, S.R. *Biochem. Biophys. Acta.* 2002. **1587** (2002), 75-82.
24. Hutter, M.M. *et al., Circulation.* **89** (1994), 355-60.

25. Kloner, R.A. and Jennings, R.B. *Circulation.* **104** (2001), 2981-9.

26. Kumar, D. and Jugdutt, B.I. *J. Lab Clin. Med.* **142** (2003), 288-97.

27. Lakatta, E.G. and Sollott, S.J. *Comp. Biochem. Physiol. A Mol. Integr. Physiol.* **132** (2002), 699-721.

28. Lennon, S.L. *et al., Acta. Physiol. Scand.* **181** (2004a).

29. Lennon, S.L. *et al., J. Appl. Physiol.* **96** (2004b), 1299-305.

30. Lennon, S.L. *et al., Am. J. Physiol. Heart Circ. Physiol.* **287** (2004c), H975-80.

31. Lesnefsky, E.J. *et al., J. Lab. Clin. Med.* **124** (1994), 843-51.

32. Liu, P. *et al., Cardiovasc. Res.* **56** (2002) 443-53.

33. Lucchesi, B., *Heart Physiology and Pathophysiology*, ed. N. Sperelakis, (Academic Press: San Diego, 2001), 1181-1210.

34. McCully, J.D. *et al., Physiol. Heart Circ. Physiol. Injury.* **286** (2004), H1923-35.

35. Pepe, S., *Clin. Exp. Pharmacol. Physiol.* **27** (2000), 745-50.

36. Powers, S.K. *et al., Am. J. Physiol.* **265** (1993), H2094-8.

37. Powers, S.K. *et al., Am. J. Physiol.* **275** (1998), R1468-77.

38. Powers, S.K. and Lennon, S.L. *Proc. Nutr. Soc.* **58** (1999), 1025-33.

39. Powers, S.K. *et al., Ann. N Y Acad. Sci.* **1019** (2004), 462-70.

40. Quindry, J. *et al., Exp. Gerontol.* (2005), *In Review.*

41. Rebrin, I. *et al., Free Radic. Biol. Med.* **35** (2003), 626-35.

42. Scarabelli, T. *et al., Circulation.* **104** (2001), 253-6.

43. Starnes, J.W. *et al., Am. J. Physiol. Heart Circ. Physiol.* **285** (2003) H347-51.

44. Tani, M. *et al., J. Mol. Cell. Cardiol.* **29** (1997), 3081-9.

45. Taylor, R.P. *et al., Am. J. Physiol.* **276** (1999), H1098-102.

46. Vetterlein, F. *et al., Am. J. Physiol. Heart Circ. Physiol.* **285** (2003), H755-65.

47. Yamashita, N. *et al., J. Exp. Med.* **189** (1999), 1699-706.

48. Yu, B.P. *Physiol. Rev.* **74** (1994), 139-62.

CHAPTER 9

GENETIC EXPRESSIONS: OXIDATIVE STRESS, EXERCISE, AND AGING

N.B. Schweitzer and H.M. Alessio
Miami University, Oxford, OH

9.1 Introduction

The interaction between genetics and environment is a major determinant of health, aging, and exercise performance (Simopoulos, 1997; Skinner *et al.*, 2003). While genes provide a blueprint for a variety of phenotypes, environmental and lifestyle factors can modify specific structural, functional, or behavioral characteristics of an organism (Fig. 9.1). Access to physical activity, in particular, is an important factor that directly influences genes that regulate health and longevity. On the other hand, chronic physical inactivity can influence genes that impair health and contribute to hypokinetic diseases. Recently there has been a growing interest in documenting gene expressions associated with exercise in order to understand underlying mechanisms by which

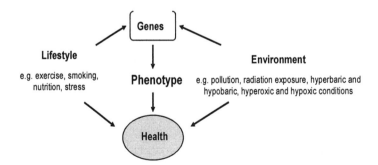

Fig. 9.1. The complex interaction of genes, exercise, and environmental factors on overall health. (adapted in part from Bray, 2000).

exercise contributes to health and life span. Identifying the timing of turning on or off specific genes that appear to be sensitive to exercise - or lack of exercise — may provide useful information for disease prevention and intervention. Specific genes such as *Edn-1, Brca1, Brca2*, and variants of the *ApoE* gene are known to increase the risk of diseases such as hypertension, breast cancer, and Alzheimers. It is not known what impact exercise or physical inactivity may have on the timing and expression levels of these genes, but there is significant interest in identifying possible interactions that impact health and disease risk.

Longevity is regulated by the interaction between genetics and the environment. This is apparent when comparing the shapes of different human population survival curves at different points in history. A cohort of individuals may have similar longevity determining genes, but environmental forces-negative or positive-may interact with genetics and result in differences in survival within the same cohort and between different cohorts. Figure 9.2 compares survival as a function of age in humans exposed to different environmental hazards specific to historical times separated by centuries. The survival curve shapes change from an exponential slope to a curvilinear to a rectangular shape. These changes relate to an environment with severe hazards to a gradual removal of environmental hazards, to advances in medical science, to the theoretical addition of a change towards a healthy lifestyle (Buskirk and Hodgson, 1990). In the rectangular-shaped survival curve of 30 years ago, causes of death are likely to have a strong genetic influence since death from environmental hazards is for the most part, removed. The theoretical survival curve is likely to reflect both genetic influences and the interaction between lifestyle and environmental forces on life span.

The graph also shows that there is a much greater difference between the average life spans of people living 50,000 years ago and today. This suggests that humans may share common longevity determining genes, but factors other than genetics interact and modify average life span. Factors such as environment, access to health care, antibiotics, advances in medicine, sanitation, psychological factors, access to physical activity, exercise, smoking, nutrition, and other lifestyle variables have had a significant influence on human survival, and specifically mean life span, over the centuries.

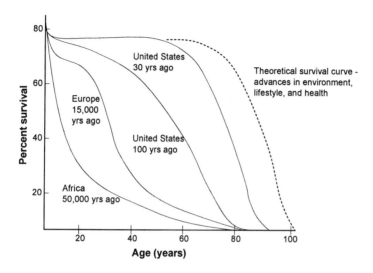

Fig. 9.2. Human survival curves evolved from random (exponential shape) to predicted (near-rectangular) and a predicted (rectangular shape) from modifications in environment, lifestyle, and health (adapted from Cutler, 1984).

It is well known that many animal species share similar genetics, biology, anatomy, and biochemistry. For example, humans share 99.4% of genes with chimpanzees and 70% with fruit flies. These shared features have assisted our understanding of the biological roles of the number and type of genes as well as the timing of when genes are expressed since short-lived animals can be studied more easily in longitudinal-designed investigations. Estimates of the total number of genes in most primates, including humans, range from 18,000 to 40,000. Despite the large number of genes and the potential that thousands of genes may be involved in regulating enzymes that catalyze specific biological reactions, most scientists agree that only a small number of genes are directly involved in controlling important biological processes related to health and aging (Finch, 1990). Approximately 3% of identified genes are thought to be involved in cell function while the other 97% have unknown function and may actually contain historical or evolutionary information (Hampton, 1991).

Two types of genes, structural and regulatory, are believed to regulate health and longevity in animals (Cutler, 1984). Structural genes determine morphology, general biology, and functioning. Regulatory genes work in partnership with structural genes to determine when they express their gene products. Most evolutionary changes that distinguish animal species are believed to occur by different timing and amount of expression of the same structural genes, not as a result of differences, changes or mutations in structural genes. For example, regulatory genes are responsible for higher amounts of antioxidants found in long-lived compared with short-lived animals. That is because despite a large range of maximum life spans, animals share virtually the same exact types of antioxidants-they just have varying amounts expressed in tissues at different times. Exercise probably influences regulatory gene expressions in many tissues over an organism's life span. This will have a major impact on health and longevity.

9.2 Gene Expression and Aging

It is estimated that 25% of the variation in life span is genetically influenced with the remainder due to environment and lifestyle (Mitchell *et al.*, 2001). Older organisms act and function in ways that are more similar than different compared with younger organisms. Centenarians, in particular, appear to have different gene expressions compared with individuals who die long before their 100th birthday. Centenarians rarely have mutations in the tumor suppressor genes *Brca-1* and *Brca-2* associated with breast and ovarian cancer (King *et al.*, 2003), and are more likely to have mutations in the cholesteryl ester transfer protein gene that reduces risk for cardiovascular disease (Barzilai *et al.*, 2003). Thus aging is in part a series of time-related, genetically controlled processes that gradually disturbs homeostasis and steers an organism towards death. Browner *et al.* (2004) reviewed genes and genetic pathways in a variety of animal models that may be implicated in human aging.

Environmental factors interact with the programmed component of aging. Table 9.1 summarizes several examples of candidate genes involved in human aging that may be modifiable by lifestyle factors, such as nutrition and exercise. Genes as diverse as those encoding antioxidant enzymes such as superoxide dismutase (SOD) and genes encoding deacetylase enzymes that silence some potentially harmful genes (e.g. cis-prenyltransferase, *Srt1* and *Srt2,* which are involved in protein glycosylation) increase due to exercise and caloric restriction. Genes for insulin and insulin like growth factor (IGF), macrophage inhibitory factor (MIF), interleukin 6 (IL-6), and C-reactive protein (CRP) decrease in individuals who exercise regularly (Fischer *et al.,* 2004; Hammett *et al.,* 2004).

Table 9.1. Candidate genes involved in human aging that are modified by lifestyle.

Candidate genes	Gene Pathways	Possible Chronic Exercise Effects
SOD	Oxidative stress	Upregulation
SRT1, SRT2	Nutrition and Caloric restriction	Upregulation
IGF	Stress resistance and insulin signaling	Downregulation
MIF, IL-6, CRP	Inflammation	Downregulation

9.3 Factors Involved in Aging and Cell Death

Growth, development and aging are processes that required cell differentiation and development. Accompanying these processes are changes in metabolism, changes in hormone levels, and changes in exposure to oxidative stress (Frolkis, 1993). It is simplistic to conclude that positive biological changes only occur before maturity and negative changes only occur after maturity, as aging is the accumulation of all processes occurring over the lifetime of the organism.

A pleiotropic condition exists in which genotypes that offer advantages at one point in time may have deleterious effects at another point in time (Finch, 1990). Compared with short-lived species, long-lived species have a longer developmental period before maturation which may provide a window of opportunity for beneficial biological actions after which time the same biological actions may become harmful. For example, a high metabolic rate may be beneficial early in life but may later lead to metabolic by-product accumulation and accompanying damage.

Prior to maturation most physiological changes are beneficial and bring about growth and development. After maturation, most physiological changes cause cell damage, loss of cell function, and cell death. Cell death can occur through both apoptosis and necrosis (Fig. 9.3). Apoptosis is a necessary self-destruction process that an organism uses to maintain homeostasis by eliminating aged, infected, or mutated cells. Necrosis is primarily due to disease or injury and unlike apoptosis, the cell loses ATP, swells, explodes, and causes inflammation that affects cells nearby.

There is a lot of interest in identifying cell signaling and genetic factors that regulate cell death. Genes such as *Bax, Fas*, and *p53* express proteins that initiate apoptosis while the proteins genes such as *Bcl-2* and *Bcl-XL* express proteins that inhibit apoptosis (Phaneuf and Leeuwenburgh, 2001). Cell death depends on the ratio of the expression of these proteins, so for example high levels of the protein BAX relative to BCL-2 will promote cell death by apoptosis (Oltvai *et al.*, 1993). It is interesting that extremes of activity, either high intensity exercise or complete immobilization, may initiate apoptosis in lymphocytes and muscle (Abravaya *et al.*, 1992) due to increases in ROS, intracellular calcium, catecholamine secretion, tumor necrosis factor and lower BCL-2 levels, all of which can signal the onset of apoptosis (Duke *et al.*, 1996; Green *et al.*, 1998). Normally, apoptosis results in a somewhat orderly removal of damaged cell fragments that will enable cell turnover and restore function. However, uncontrolled apoptosis can lead to a variety of age-related diseases including heart disease, autoimmune deficiency syndrome and Alzheimers's disease.

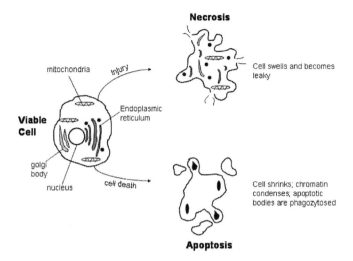

Fig. 9.3. After maturation, most physiological changes may cause cell apoptosis or necrosis.

9.4 Skeletal Muscle Aging

Damaged DNA and proteins are characteristic of aging (Frolkis, 1993). Nevertheless, many cells including those that make up skeletal muscle tissue are extremely resilient over time. This was demonstrated in a study by Jozsi *et al.* (2000). Skeletal muscle gene expressions of 60 and 70 year old men were compared with 20 and 30 year old men. There were striking similarities between the healthy young and old muscle. The gene analysis showed less than 2% difference in all genes in subjects of all ages. This finding indicates that healthy muscle is remarkably stable with age. Older muscle had 2-4 fold higher expression of only six genes that represented muscle damage and only one gene that expressed for DNA repair. These specific genes, their functions, and age differences are shown in Table 9.2.

Table 9.2. Genes, function, and age changes in young and old healthy men (Jozsi *et al.*, 2000).

Genes	Function	Age Change (%)
Heat shock protein 27	Actin-cap binding protein that contributes to cytoskeltal reorganization	+ 170
Natural killer cell enhancing factor	Regulates antioxidant proteins	+ 180
MAP kinase kinase 3	Mediates p38 MAP kinase signaling pathway and increases expression of proinflammatory cytokines for cell repair	+ 280
Thymosin beta-10	Actin-binding protein with important roles in neural growth, cytoskeletal organization, and apoptosis	+ 250
Cellular nucleic acid binding protein	Transcription factor	+ 210
X-ray repair cross complementing	DNA repair base excision protein gene	- 70

When comparing some of the differences in gene expression in muscle tissue from different aged healthy men and young muscle, old muscle appears to be in a state of stress based on the following characteristics: (1) inflammation (280%), (2) elevated cytoskeletal organization (170%), (3) neural growth (250%), (4) elevated antioxidant activity (180%), and (5) apoptosis (250%). Inflammation was observed through increased expression of the *Mkk3* gene, which is responsible for activating the p38 MAP kinase pathway. This pathway is essential for muscle repair signaling via proinflammatory cytokines (Jozsi *et al.*, 2002). In addition, elevated cytoskeletal reorganization and neural growth occur as muscles respond and adapt to novel or stressful stimuli. Elevated antioxidant activity in old muscle may represent an increased need and attempt to protect DNA from oxidative stress. The lower (-70%) X-ray repair cross complementing gene expression in older muscle may represent a decline in DNA repair capacity despite the elevation of other genes related to cell structure, cell signaling, and antioxidant protection. Based upon these differences between old and young skeletal muscle, Jozsi *et al.* (2000) concluded that types and levels of gene expression in aging muscle demonstrated evidence of stress and damage. Altered expression of just a few genes in the older muscle lowered its ability to effectively respond to exercise, making it more susceptible to exercise-induced oxidative stress.

9.5 Oxidative Stress and Gene Regulation

Genetic control of antioxidant levels and prooxidant systems show evidence of both long-term and immediate actions. Both systems can adapt within minutes to environmental stress. One of the most powerful stressors for inducing changes in skeletal muscle phenotype, is exercise-induced oxidative stress (Hargreaves and Cameron-Smith, 2002). Muscles can adapt to exercise in a variety of ways: (1) metabolic shifts to increase energy transfer, (2) hypertrophy, (3) repair of muscle damage, (4) induction of antioxidant genes, and (5) repression of oxidases that increase prooxidants.

9.6 Reactive Oxygen Species Control Gene Expression

Exercise regulates gene expression via reactive oxygen species (ROS) that are produced by oxidative stress reactions. Oxidative stress can be elevated in different ways by different types of exercise. Some types of exercise increase mechanical stress in muscles while other types increase metabolic stress. Both can result in the increased production of ROS. ROS influence gene expression by oxidizing and modifying DNA or by acting as cell signaling molecules that turn on or off certain genes that express for specific pro and antioxidant molecules. A GenMAPP formulated by Reyes and Reyes (1999) describes oxidative stress-regulated pathways that inhibit specific prooxidant pathways and induce antioxidant pathways. For example, ROS inhibit expression of genes for NADPH oxidase, xanthine oxidase, and monamine oxidase. On the antioxidant side, JNK3 and MAPK p38 initiate a cascade of regulatory factors controlling expression of antioxidants including SOD and catalase (Fig. 9.4). In addition to antioxidants, NF-αB transcription factors also regulate immune response, cell apoptosis, inflammation, oncogenesis, and various diseases. The most common and best characterized form of NF-αB is the p65-p50 heterodimer. It is activated by a variety of stressors including growth factors, cytokines, tumor necrosis factor-α (TNFα) and exercise. NF-αB plays a role in regulating virtually every possible type of cell in the body.

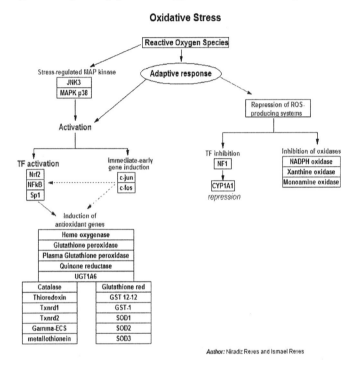

Fig. 9.4. Oxidative stress pathway from Reyes and Reyes (1999).

Several transcription factors that specifically sense ROS have been identified. When Oxy R is oxidized by a hydroxyl radical, it undergoes a conformational change increasing its ability to bind to promoters and in turn upregulate expression of its target gene. Sox R also increases gene transcription with elevated affinity of the promoter to RNA polymerase, but unlike Oxy R, Sox R is activated by superoxide radicals. When exposed to oxidative stress, the eukaryotic transcription factor Skn-7 regulates transcription of antioxidant proteins, such as glutathione reductase. Additional involved transcription factors may trigger the transcription of stress genes that act to resolve cellular inflammation and therefore return oxidative stress back to a manageable level (Lane *et al.*, 2003). Lane *et al.* (2003) described a mechanism whereby old cells lose the capability to reduce inflammation or disease through the initiation of transcription factors for stress reduction genes, and therefore the chronic inflammatory response disturbs gene regulation and expression.

9.7 Advances in Gene Investigation

Genes are units of heredity information found on the chromosomes in the cell nucleus. They consist of sequences of nucleotides that encode various proteins which are expressed via transcription and translation (Fig. 9.5). Genes specify the genotype of an organism, while the characteristics that arise from the interaction of an organism's genes and environment become an organism's phenotype. The turning on or off of specific genes is regulated by an array of causes including transcription factors, hormones, cell temperature (Singh, 2003), and histone methylation which opens up the chromatin so other proteins can access the gene (Strahl *et al.*, 2004; Zhang *et al.*, 2003). The type and amount of gene expression provide insight into the up- or downregulation of specific genes that encode for proteins that influence cell function and survival. There is growing interest in documenting the influence of exercise on specific gene expressions and then relating specific gene expressions to biological outcomes associated to disease risk factors such as high blood pressure and cholesterol.

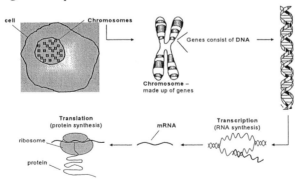

Fig. 9.5. Chromosomes containing DNA undergo transcription for RNA synthesis and translation for protein synthesis.

9.8 Gene Expression and Environment

Exercise, aging, and oxidative stress exhibit a feedback relationship with gene expression: each condition can influence gene expression and gene expression can influence each condition. Recent advances in studying the genome have provided more clues than ever before about the way specific gene expression can regulate phenotypic characteristics of organisms associated with health and aging. Advances in DNA analysis have moved us from using a single cell gel electrophoresis such as comet assays to measure DNA damage in individual cells to northern blots and reverse transcriptase polymerase chain reaction (RT-PCR) to detect and quantify mRNA. DNA microarray analysis can provide information on tens of thousands of genes in a single experiment. From this profile, genes that are up- and downregulated can be distinguished, gene maps can be produced, and specific genes that could be responsible for certain diseases or specific biological processes associated with exercise, aging, and oxidative stress can be identified.

Transcriptional profiles of essentially the entire genome—including all of the expressed traits—can be compared with a variety of health and disease conditions (Kaminski and Friedman, 2002). Although the microarray technology takes a tremendous leap past previous gene analysis technology that was limited to examining one gene or a small cluster of genes at a time, inconsistencies can still be found with this technique. It is becoming a standard laboratory method to validate the microarray results through the amplification of DNA by quantitative RT-PCR or immunohistochemistry.

9.9 High Density Oligonucleotide Arrays

The emergence of the high-density oligonucleotide array, or microarray, technology revolutionized the ability of scientists to uncover novel genes and variations between gene expressions. A decade ago, the concept of simultaneously measuring the expression of thousands of genes in a single experiment was implausible. Today, the knowledge of gene sequencing has empowered the advancement of genome technology and the simultaneous profiling of the expression of tens of thousands genes with microarray technology is becoming a standard laboratory technique (Schena *et al.*, 1996; Tavazoie *et al.*, 1999; Lockhart and Winzeler, 2000). Importantly, the use of microarray technology gives researchers the ability to infer functions for newly identified genes based on the expression patterns of known genes. Understanding the relationship between gene families enables researches to identify and characterize genes at the cellular level, potentially those genes involved in disease.

Among the different types of gene expression arrays, Affymetrix GeneChip arrays are the most popular arrays used in biomedical research. The microarray uses a set of DNA probes to measure specific gene expression. Normally, each gene has pairs of related oligonucleotides known as a probeset (Liu *et al.*, 2003). Each probe pair consists of a perfect match probe and a mismatch probe (Roy *et al.*, 2002). The microarray uses mRNA from the biological sample of interest as a template for cDNA synthesis. The cDNA can be used as a template for synthesis of cRNA using fluorescently labeled base analogs. The microarray is then exposed to the cRNA, which binds, or hybridizes, to complementary sequences on the microarray (Fig. 9.6). Arrays are scanned and analyzed for intensity values of each probe. The intensity is correlated to the amount of hybridized cRNA and therefore to the degree of gene expression (Fig. 9.7). Direct comparison of gene expression in a treatment and a control group is possible, and both well-characterized and unknown gene expression can be assessed.

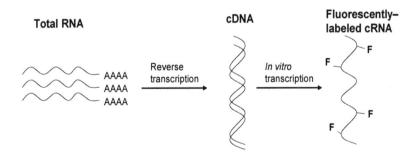

Fig. 9.6. To detect cRNA bound to a microarray, it must be labeled with a reporter molecule, such as a fluorescent tag.

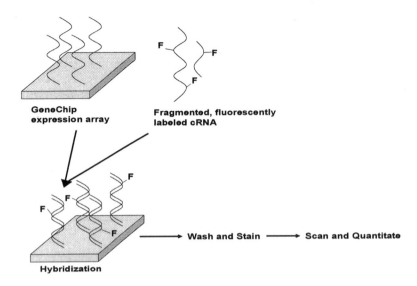

Fig. 9.7. Fluorescently labeled cRNA will hybridize with different DNA sequences on the GeneChip expression array allowing for quantification of gene expression.

9.10 Gene Expression and Exercise

Exercise and physical activity may be one of the most important modifiable environmental factors related to health (Chen, 2001). Using microarrays in exercise physiology will provide new insights into the complex mechanisms of the exercise-induced gene expression responses (Table 9.3). Analyzing gene expressions will likely characterize and predict individually variable responses to exercise. For example, acute exercise usually results in elevated genes expressions for stress, cell remodeling, and inflammation. Genes such as *Il-1B* encode inflammatory cytokines that stimulate neutrophil action and initiate cell repair process (Jozsi *et al.*, 2000). Genes such as *Vegf* play a major role in angiogenesis, revascularization, and production of vasoactive molecules (Richardson *et al.*, 1999). Expression of the gene for *c-Jun* also increases following exercise and this facilitates growth, differentiation, and apoptosis as well as denervation and reinnervation of muscle fibers (Herdegen *et al.*, 1997). In general, acute exercise modifies gene expressions for removing damaged cells and activating inflammatory and immune responses so that repair of cell damage can occur quickly followed by remodeling and in some cases, hypertrophy.

Table 9.3. Examples of microarrays used in exercise studies to determine exercise-induced gene expression.

	Gene Expression Array	**Gene Expression Profile**	**Gene Function**
Diffee *et al.* (2003)	Affymetrix Rat Genome U34A array	Increase in atrial MLC-1 in exercised rat ventricles	Myocardial contraction
Tong *et al.* (2001)	Affymetrix Rat Genome U34A array	Increase in BDNF in exercised animals	Neuronal survival and plasticity
Zambon *et al.* (2003)	Affymetrix Test2 array	Resistive exercise regulated Per2, Cry1, and Bmal1	Circadian clock regulation

A recent study by Schweitzer *et al.* (2005) used the Affymetrix® DNA microarray analysis to obtain data about genes expression associated with different access to exercise. Physiological measurements (e.g. systolic blood pressure, diastolic blood pressure, heart rate, cholesterol and triglycerides) and genes associated with cardiovascular (CV) regulation were compared among animals having: (1) no access to physical activity (2) access to hour-long, twice-weekly activity in a large box, and (3) access to a running wheel. The amount of activity differed among the three groups. Animals with access to a running wheel ran 5642 ± 1773 m•day^{-1} at 2 months of age and gradually declined, ranging between 1612 and 4836 m•day^{-1} throughout the remaining of the study. These distances were dramatically different than the 129 m•day^{-1}covered by sedentary animals residing in a standard cage. Animals with access to twice weekly physical activity covered a distance ranging from 161 and 322 m•day^{-1}depending on whether they were monitored in light or darkness. This distance was still a fraction of what was covered in the running wheel group. The animals in this study proved an interesting model when evaluating the importance of environment on genetic expression. The normal environment in a laboratory experiment limits the amount of physical activity or exercise animals receive. This sedentary environment may have detrimental effects on genes that impact health and longevity.

Schweitzer *et al.* (2005) found exercise-induced gene expression changes among 16 month old rats that were similar to results of Jozsi *et al.* (2002), although the percent fold changes were smaller, possibly due to the different tissues (heart vs. skeletal muscle) and gene analyses techniques. Schweitzer *et al.* (2005) found the mean gene expressions for *Gja1, Fdft1, Edn1, Cd36,* and *Hmgb2* expressed by the left ventricle to be different among animals that had access to physical activity compared with animals that resided solely in a standard cage (Fig. 9.8). Farnesyl-diphosphate farnesyltransferase *(Fdft1),* located on chromosome 15p16alt.16, is associated with cholesterol biosynthesis and was increased by 11% in the sedentary animals compared to exercised animals. Gap junction protein alpha 1 *(Gja1),* located on chromosome 20q11-q13, is an integral component of heart rate and was upregulated by 22% in the sedentary group. A lower expression (29%) of Endothelin-1 *(Edn1),* located on chromosome 17, found in exercise animals corresponds to relaxed, dilated vasotone and plays a role in maintaining a low systolic blood pressure. *Cd36,* located on chromosome 4, is associated with fatty acid homeostasis and cell adhesion and had 10% lower expression in exercised compared to sedentary animals. High mobility group box 2 *(Hmbg2),* located on chromosome 16p14-q11, is involved in regulation of transcription and neurogenesis and was regulated by 12% in sedentary animals. The different expression of these genes, in addition to physiological health markers, suggested that an animal's phenotype can be genetically altered and influenced by the exercise.

The Heritage Family Study, funded in 1992 by the National Institutes of Health has followed 800 Eurocaucasian and African American men and women from over 200 families. Generally, the results from the Heritage study have shown that genetics play a significant role in differentiating responders and non-responders to similar exercise regimes and mechanisms that explain how exercise may impact the risk of many different types of diseases. One of the many studies that compared genetic factors that contribute to skeletal muscle phenotypes reported a surprisingly weak familial resemblance for muscle fiber types before and in response to regular exercise training. On the other hand, maximal enzyme activities for major metabolic pathways were influenced by genetic factors (Rico-Sanz, 2003). A comparison of familial resemblance for coronary heart disease showed that coronary heart disease runs along family lines, and common environmental effects were important in explaining the observed familial resemblance (Katzmarzyk *et al.*, 2000).

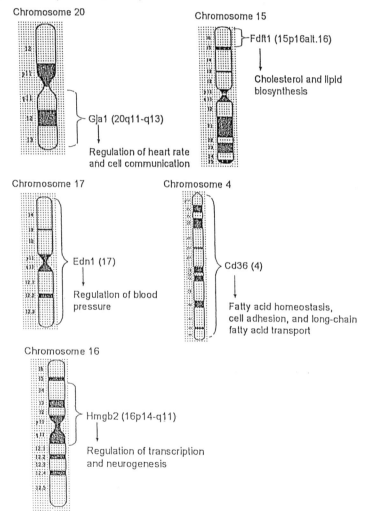

Fig. 9.8. *Gja1, Fdft1, Edn1, Cd36*, and *Hmgb2* which are located on different chromosomes, differed between exercised animals and sedentary animals.

9.11 Exercise and Lifespan

Exercise performed on a regular basis may influence gene expressions that regulate health and lifespan. Several large scale studies have documented increases in mean life span associated with regular exercise in both healthy (Sesso *et al.*, 2000; Blair *et al.*, 1996) and diseased (Leon *et al.*, 1996) populations. Increased life years gained from regular exercise range from 2-7 and are thought to be the result of physiological changes to cardiorespiratory and immune function (Sesso *et al.*, 1996) and enhanced ability to protect against oxidative stress and chronic diseases including cancer (Lee *et al.*, 1994). Recently, the capacity for life-long exercise to either up- or down-regulate certain genes was investigated by Bronikowski *et al.* (2002). In order to study the effects of voluntary life-long exercise on life span, Bronikowski *et al.* (2002) investigated active animals that were chosen from a genetically selected colony of running mice. Selectively bred active animals, which voluntarily run more than twice the distance of control animals, were divided into either a forced sedentary environment or an environment with access to a running wheel.

Compared with a sedentary group, median lifespan, but not maximum lifespan, was 17 % higher in the exercise group (Fig. 9.9). Active animals had a median lifespan of 698 days while the median sedentary lifespan was 599 days. Physical activity appears to have a similar effect on life span as environmental factors did as shown in human survival curves from 50,000 and 30 years ago. That is, more animals live longer when they have access to and participate in life-long physical activity. But maximal life span appears to be fixed. It is unaffected by either the removal of major environmental hazards as shown in human survival curves and is unaffected by life-long physical activity as shown in animal survival curves (Bronikowski *et al.*, 2002; Holloszy, 1995).

9.12 Gene Expression in Sedentary Aged Animals

It may seem intuitive that compared with physically active animals, sedentary animals would in general, have lower gene expressions, perhaps due to less mechanical or metabolic stimulation. Interestingly, Bronikowski *et al.* (2002) reported that microarray analyses showed sedentary animals had a greater number of gene expressions compared with physically active animals. They reported a general decrease in the number of deleterious age-related gene expressions in the exercise group compared with the sedentary group. In the physically active group, fewer genes involved in inflammatory response, stress response, signal transduction, and energy metabolism were significantly altered with age compared to the sedentary population. This suggests that compared with sedentary animals, physically active animals experienced more stability and less cell damage over time. Bronikowski *et al.* (2002) found inflammatory and stress response genes were the most affected by age within the sedentary group. Twenty-one inflammatory response genes were up-regulated with aging in the sedentary group compared to only 9 in the active group. For example, the authors found an increase in complement genes, such as *C1qb, C1qc,* and *Complement C4,* which are involved with immune function. The low level of expression of these genes in the physically active animals suggests that life-long exercise minimized or even prevented tissue inflammation.

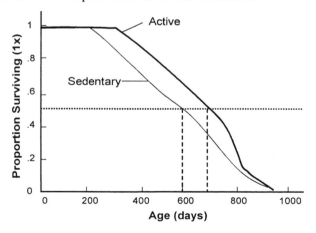

Fig. 9.9. Survival curves for active and sedentary animals. Median lifespan was greater in active animals compared to sedentary, 698 vs. 599, respectively (adapted from Bronikowski *et al.*, 2002).

9.13 Oxidative Stress in Genes: A Shared Mechanism in Exercise and Aging

It is widely accepted that reactive oxygen (ROS) and nitrogen (RNS) species are key influential factors in aging and disease processes. Radical production from either oxygen or nitrogen-centered molecules occurs as a result of normal metabolic reactions as well as abnormal or reactive responses to cellular attack by microbes or other foreign species. Radical molecules are likely to play a role in cell breakdown because they are able to alter and inactivate enzyme complexes, damage DNA and RNA, and alter membranes. The longer the cell lives, the more radicals are produced and the more likely these radicals will attack vital cell components, including genes. So, with age, genetic damage is a real threat. The most susceptible components of DNA to radical attack are pyrimidines, purines, and deoxyribose (Saul *et al.*, 1987). The hydroxyl radical is a highly reactive oxygen-centered radical that will react with all molecules in living cells and modify the pyrimidine and purine bases (Halliwell, 1994) (Fig. 9.10).

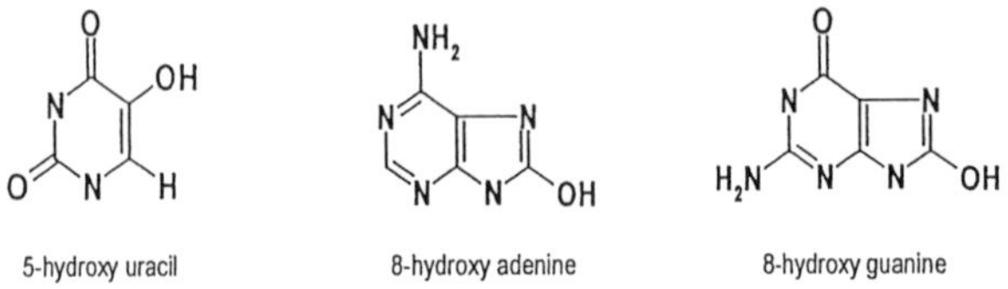

Fig. 9.10. Purine and pyrimidine bases of DNA are subject to attack by hydroxyl radicals.

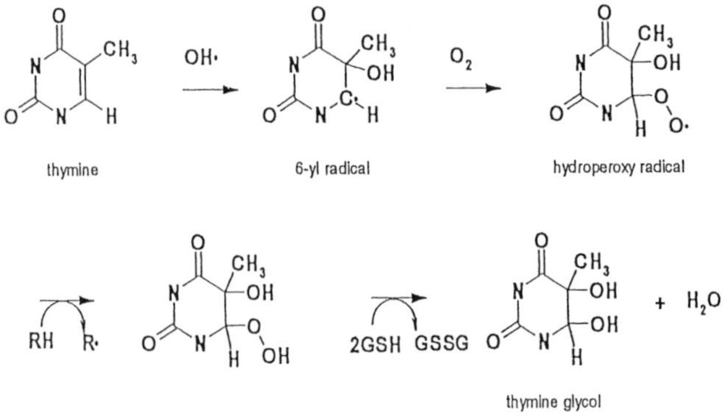

Fig. 9.11. Mechanism for the formation of thymine glycol from hydroxyl radical addition to thymine (adapted from Saul *et al.*, 1987).

For example, when thymine is attacked by a hydroxyl radical, an organic hydroperoxide is formed and will undergo a chemical reduction to form the alcohol, thymine glycol (Fig. 9.11). Thymine glycol is formed by either oxidative stress or by oxidation of the 5,6 double bond of thymine when exposed to radiation.

Incorporation of oxidized based into newly synthesized DNA is prevented by a system of enzymes such as the human MutT homologue, *hMTH1*. An inverse correlation between oxidized guanosine and *hMTH1* has been reported by Sato *et al.* (2003) suggesting that *hMTH1* is effective in protecting DNA.

Structural changes to pre-existing DNA have been linked to aging and various diseases including cancer (Ames *et al.*, 1993). DNA damage like that described above can be repaired by: (1) repair of single strand breaks, (2) repair of double strand breaks, (3) nucleotide excision repair, (4) base excision repair, (5) photreactivation, and (6) postreplicative repair. Of these six possible ways, excision-repair was probably the first mechanism that evolved in eukaryotic organisms to replace DNA damaged by radiation or oxygen-induced radical reactions. Two different types of excision-repair systems exist in the nucleus that can restore radical damage to DNA: (1) a general and (2) specific repair system (Saul and Ames, 1990). The general repair mechanism involves an excision exonuclease enzyme that removes an entire portion of DNA that contains the damaged base. A more efficient method is by the DNA glycosylase enzyme that removes only the specific base that was damaged by radicals (Fig. 9.12). In both processes, most of the oxidized (e.g. damaged) bases removed from the cell are then excreted in the urine. An example of an oxidized base that has been recovered in urine is 8-hydroxyguanine. This mutagenic lesion is produced when guanine is oxidized by hydroxyl radical. Formation of 8-hydroxyguanine halts DNA replication until repair processes are set into motion.

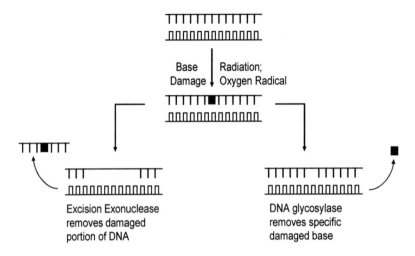

Fig. 9.12. Excision-Repair.

Both the general excision exonuclease and the specific DNA glycosylase methods are important for survival as the cell can replace or repair damaged bases before an error in RNA or protein synthesis occurs. Either type of error would have long-lasting effects. Errors in repair can lead to mutations or changes in DNA. It is estimated that a single cell endures over 100,000 oxidative "hits" each day. DNA is constantly under siege and undergoes continuous oxidative damage and repair processes.

High-intensity exercise damages DNA based upon elevated urinary 8-OHdG levels following exercise (Radak *et al.*, 2003). Low-intensity exercise, on the other hand does not result in increased 8-OHdG levels, except in sedentary individuals (Sato *et al.*, 2003). They also reported that compared with sedentary individuals, trained individuals had higher expression of *hMtH1* mRNA and lower levels of 8-OhdG following low intensity exercise. Most studies agree that the extent of DNA damage with exercise depends on exercise intensity (ESSR, 2003). 8-OHdG has not been a reliable indicator of aging, however, that may be partly due to its non-linear accumulation over time (Kaneko *et al.*, 1996) with the most dramatic increase occurring in old age.

It might seem paradoxical that exercise-induced oxidative stress might cause DNA damage that could lead to increased levels of oxidized bases such as 8-OHdG and possible mutations and risk for diseases such as cancer. However, in trained individuals three putative adaptations occur which work together to reduce the amount of oxidative DNA damage and facilitate repair: (1) decreased ROS production especially during submaximal exercise, (2) enhanced antioxidant defense system, and (3) enhanced DNA repair.

9.14 Summary

Environmental factors (e.g. physical activity access, time, diet, and geography) influence genetics in a dynamic process in which environmental forces can modify specific characteristics of an organism. Access to physical activity, in particular, is an important environmental factor that directly influences genes that regulate health and longevity. Contrarily, physical inactivity can influence genes that impair health and contribute to disease by damage to DNA or proteins. Therefore it is this interaction of exercise on gene expression that fuels the interest in documenting gene expressions associated with exercise, health and disease. Over the past ten years, gene expression analysis has made tremendous leaps, making it possible for scientists to compare genes across an entire genome and eventually facilitating the exploration of unknown genes.

References

1. Abravaya, K. *et al., Genes and Dev.* 6 (1992), 1153-64.
2. Anderson, G.J. *et al., Physiol. Behav.* 70 (2000), 425-429.
3. Barzilai *et al., JAMA.* 290 (2003), 2030-2040.
4. Blair, S.N. *et al., JAMA,* 276 (1996), 205-10.
5. Bray, M.S. *J. Appl. Physiol.,* 88 (2000), 788-792.
6. Bronikowski, A.M. *et al., Physiol. Genomics.* 12 (2003), 129-138.
7. Buskirk, E.R. and Hodgson, J.L. *Fed.Proceed.* 46 (1987), 1824-1829.
8. Chen, J., *World Rev. Nutr. Diet.* 89 (2001), 108-117.
9. Cutler, R.G. *Aging and Cell Function,* ed. J.E. Johns, Jr (New York: Plenum, 1984), 1-148.
10. Diffee, G.M. *et al., Am. J. of Heart Circ. Physiol.* 284 (2003), H830-H837.

11. Duke, R.C. *et al.*, *Scientific American.* **275** (1996), 80-7.
12. Finch, C.E. *Longevity, Senescence, and the Genome.* (University of Chicago Press, Chicago, 1990), 42-59.
13. Fischer *et al.*, *Am.J. Physiol.* **287** (2004), E1189-1194.
14. Frolkis, V.V., *Mech. Ageing Dev.* **69** (1993), 97-107.
15. Green, D.R. *Cell* **94** (1998), 695-8.
16. Hammett, C.J.K. *et al.*, *J Am Coll Cardiol.* **44** (2004), 2411-2413.
17. Hampton, J.K. *The biology of human aging* (William C. Brown Publishing, Dubuque, IA, 1991).
18. Hargreaves, M. and Cameron-Smith, D. *Med. Sci. Sports Exerc.* **2** (2002), 1505-1508.
19. Halliwell, B.H. *Ann. Neurol.* **32** (1992), 510-515.
20. Herdegen, T. *et al.*, *Trends Neurosci.* **20** (1997), 227-231.
21. Jozsi, A.C. *et al.*, *Mech. Ageing Dev.* **120** (2000), 45-56.
22. Kaminski, N. and Friedman, N. *Am. J. Respir. Cell. Mol. Biol.* **27** (2002), 125-132.
23. Katzmarzyk, P.T. *et al.*, *Ethn Dis.* **10** (2000), 138-147.
24. King, M.C. *et al.*, *Science.* **302** (2003), 643-646.
25. Lane, N. *J. Theor. Biol.*, **225** (2003), 531-540.
26. Leon, A S. *et al.*, *J. Cardiopulm Rehab.* **16** (1996), 183-192.
27. Liu G. *et al.*, *Nucleic Acids Res.* **31**(2003), 82-86.
28. Lockhart, D.J. and Winzeler, E.A. *Nature.* **405** (2000), 827-836.
29. Mitchell, B.D. *et al.*, *Am. J. Med. Genetc.* **102** (2001), 346-352.
30. Oltvai, Z.N. *et al.*, *Cell.* **74** (1993), 609-19.
31. Sesso, H.D. *et al.*, *Circulation* **102**(2000), 975-80.
32. Lee, I.M. *Med. Sci. Sports Exerc.* **26** (1994), 831-7.
33. Phaneuf, S. and Leeuwenburgh, C. *Med.Sci. Sports Exerc.* **33** (2001), 393-396.
34. Radak, Z. *et al.*, *Free Radical Biol. Med.* **27** (1999), 69-74.
35. Radak, Z. *et al.*, *Exerc. Immunol. Rev.* **7** (2001), 90-107.
36. Reyes, M. and Reyes, I. *Biochem. J.* **342** (1999), 481-96.
37. Richardson, R.S. *et al.*, *Am. J. Physiol. Heart Circ. Physiol.* (1999), H2247-2252.
38. Rico-Sanz *et al.*, *Med.Sci.Sports Exerc.* **35** (2003), 1360-1366.
39. Roy, S. *et al.*, *Methods enzymol.* **353** (2002), 487-97.
40. Saul, R.L. *et al.*, *Modern Biological Theories of Aging.* ed. Warner, H.R. *et al.* (Raven Press, NY, 1987), 113-129.
41. Schena, M. *et al.*, *Proc. Natl. Acad. Sci.* USA, **93** (1996),10614-10619.
42. Sen, C.K. *et al.*, *J. Appl. Physiol.* **73** (1992), 1265-1272.
43. Sen, C.K. *Sports Med.* **31** (2001), 891-908.
44. Simopoulos, A.P. *Nutr. and Fitness.* **81** (1997), 61-71.
45. Singh, H. *et al.*, *J. Cell. Biochem.* **35** (2000), 61-8.
46. Skinner, J. *et al.*, *Med. Sci. Sports. Exerc.* **35** (2003), 1908-13.
47. Socci, D.J. *et al.*, *Brain Res.*, **693** (1995), 88-94.
48. Somani, S.M. *et al.*, *Pharmacol. Biochem. Behav.*, **50** (1995), 635-639.
49. Strahl, C. and Blackburn, E.H. *Nucleic Acids Res.* **22** (1994), 893-900.
50. Tavazoie, S. *et al.*, *Nat. Genet.* **22** (1999), 281-285.
51. Tong, L. *et al.*, *Neurobiol. Disease.* **8** (2001), 1046-1056.
52. Zambon, A.C. *et al.*, *Genome Biol.* **4** (2003), R61-R61.11.
53. Zhang, Y. *Genes and Dev.* **17** (2003), 2733-2740.

INDEX